CITRUS
How to Select, Grow and Enjoy

By
**Richard Ray &
Lance Walheim**

Contributing Editor: George Vashel, Menlo Growers
Research Editor: Karl Opitz
Photography: Michael Landis

HPBooks ®
Publisher: Rick Bailey
Editorial Director: Randy Summerlin
Editors: Jonathan Latimer and Scott Millard
Art Director: Don Burton
Director of Manufacturing: Anthony B. Narducci

For Horticultural Publishing:
Art Director: Richard Baker
Associate Editor: Michael MacCaskey

Copy Editor: Barbara Young; Design: David Conover, Judith Hemmerich, Betty Hunter, Nami Takashima; Illustration: Elizabeth Dudley; Additional Photography: William Aplin, Dr. L. T. Atlas, Max E. Badgley, W. P. Bitters, W. W. Jones, Paul Moore, Robert Platt, Colvin L. Randall, Ted A. Rogers, Charles Sullins, Sunkist Growers, William Talbot, Horace V. Terhune, David L. Thomas.
Cover Photo: Michael Landis

Produced by Horticultural Publishing Co., Inc.
Published by HPBooks, Inc., P.O. Box 5367, Tucson, AZ 85703 (602) 888-2150
ISBN: 0-89586-076-7
Library of Congress Catalog Card Number: 80-82383
©1980 HPBooks, Inc.   Printed in U.S.A.
3rd Printing

This book would not have been possible without the generous assistance of:

William P. Bitters, Professor of Horticulture and Horticulturist, University of California, Riverside

John Carpenter, Plant Pathologist, USDA Citrus & Date Experiment Station, Indio, California

Ray Copeland, Superintendent, University of California Lindcove Field Station, Exeter, California

Clem Meith, University of California Agricultural Extension Agent, Butte County, Retired

Paul Moore, Superintendent of Agricultural Operations, University of California, Riverside

Albert Newcomb, Citrus Propagation Consultant, Corona, California

Robert G. Platt, Extension Subtropical Horticulturist, University of California, Riverside

Barbara Robinson, Manager of Consumer Services, Sunkist Growers, Inc.

Linda Shepler, Consumer Affairs, Sunkist Growers, Inc.

We are also grateful for the help of:

Darwin Atkin, Research Associate, University of California Lindcove Field Station, Exeter, California

William D. Adams, Texas A&M Agricultural Extension Agent, Harris County

Dr. L.T. Atlas, Horticulturist, Houston, Texas

Dean Bacon, Coordinator, University of Arizona Salt River Valley Citrus Experiment Station, Tempe, Arizona

William C. Chapman, Horticulturist, League City, Texas

B.L. Childers, Horticulturist, Port Neches, Texas

Major C. Collins, Horticulturist, Tifton, Georgia

Donald Dillon, Four Winds Growers, Fremont, California

Tom W. Embleton, Professor of Horticulture and Horticulturist, University of California, Riverside

Keith French, Villa Park Orchards, Orange, California

Raul Gonzales, Principal Agricultural Technician, University of California Lindcove Field Station, Exeter, California

Eric Johnson, Landscape Consultant, Palm Springs, California

Jerry Parsons, Texas A&M Agricultural Extension Agent, Bexar County

Fred Peterson, Soil And Plant Laboratory Inc., Santa Clara, California

Lewis Robinson, Editor-Publisher Citrograph Magazine, Los Angeles, California

Ray Sadonka, Turk Hessellund Nursery, Monticeto, California

John Smith, University of California Agricultural Extension Agent, Sonoma County

J.W. Stephenson, Horticultural Consultant, San Jose, California

Lowell True, University of Arizona Agricultural Extension Agent, Maricopa County

Dowlin L. Young, Young's Nurseries, Thermal, California

The entire staff of Sunkist Growers, Inc.

Table of Contents

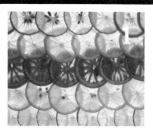

Citrus: Yesterday and Today
page 5

Citrus travels *9*, Citrus: drenched in history *10*, Citrus in the west *14*, From tree to market *18*

Citrus Climates
page 25

The world's citrus areas *26*, Florida and California *28*, California inland, coastal and desert *29*, Cold hardiness *31*, Microclimates *32*, The citrus climates of the West *35*

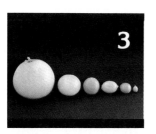

Which Varieties Will You Choose?
page 41

Oranges *45*, Oranges at the supermarket *66*, Mandarins *68*, Mandarins at the supermarket *80*, Lemons *82*, Lemons at the supermarket *87*, Grapefruit *88*, Grapefruit at the supermarket *91*, Limes *92*, Limes at the supermarket *96*, Pummelos *97*, Tangelos *102*, Tangors *106*, Kumquats *108*, Citron *114*

Growing Citrus Indoors, Outdoors, Everywhere
page 117

Soil *120*, Tree selection *121*, Planting *122*, Watering *123*, Fertilizing *124*, Pruning *128*, Growing dwarf citrus *130*, Citrus spoilers *132*, Citrus pests *134*, Growing citrus in Texas *137*, Growing citrus outside the citrus belt *140*

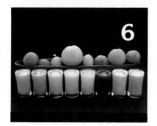

Landscaping with Citrus and Their Relatives
page 149

The citrus hedge *159*, Citrus shrubs and citrus in containers *160*, Citrus espaliers *161*

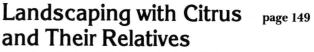

Enjoying Citrus
page 163

Juicing fresh citrus *164*, Use the whole fruit *168*, Ways to peel and slice *170*, Using citrus rinds *171*, Decorative uses of citrus *172*

Citrus: Yesterday and Today

1

There is something special about citrus. Gardeners and nongardeners alike have sensed the special qualities of citrus for generations. Citrus trees and fruits are woven into the mythologies and religions of many cultures. Eight hundred years ago Chinese agriculturists wrote of dozens of citrus varieties and their culture. Today, no fruit has been as scientifically researched and written about as citrus.

Gardeners do not just plant citrus fruits—they adopt them. When citrus enters the garden, gardeners change. They live with lemons, limes, grapefruit, oranges, tangerines and all the wonderful citrus varieties. Let us look at the life of a dwarf navel orange when it enters the garden:

In a five gallon can, it is a proud two feet in height.

It's early spring after the first flush of growth. The gardener carefully observes new growth.

Weather warms with late spring and brings the promising shower of flowers. A special fragrance is added to the garden.

Watering carefully now, perhaps misting the trees daily. The flowers drop. How many fruit set? How many will drop?

Through the summer months the fruit grows larger, and as frost nears, giant oranges, juicy, fragrant and flavorful, adorn the dwarf tree.

ORANGERIES

The attraction of citrus trees has always meant growing them in areas of unsuitable climate. The Romans wrote at length of the variety of shelters, windbreaks and other means they employed when growing the trees in northern districts. Today, innumerable citrus trees grow well beyond the traditional "citrus belt" of the world. Growers move their plants indoors in winter and provide the special care necessary. Their determination continues a long tradition.

A rainbow of sliced citrus. Top to bottom, 'Fremont' mandarin, 'Lisbon' lemon, 'Sanguinelli' blood orange, 'Meyer' lemon and 'Mexican' lime.

The orangerie of the Bagatelle, the villa and garden built for the youngest brother of Louis XVI in 1777. Earliest orangeries had at least one wall, usually south facing, made up of large windows.

In Europe in the fourteenth century, citrus culture within an artificial greenhouse climate became particularly popular. The greenhouse principle was most likely not invented for citrus, but citrus, primarily oranges and citrons, did provide the motivation that led to their wide use. Thus, greenhouses at this time were known as "orangeries."

By far the most elaborate of these orangeries was that built by Louis XIV at Versailles, France in 1682 A.D. The main body of this building was 500 feet long and 40 feet wide, supplemented by wings 350 feet long at both sides. Louis XIV used the potted citrus as ornamentals about the courtyard and in the palace, especially when entertaining at state occasions. This building still exists and is still used for this purpose.

CITRUS OF MYTH

There are many legends concerning citrus in ancient mythology. One of these suggests that at the wedding of Jupiter and Juno, the king and queen of the gods, a tree sprang up bearing golden fruit. So proud were the gods of this fruit, but so fearful of its being stolen, that they placed it on the Isles of Hesperides with Atlas the giant to guard them. The hero Hercules was assigned as one of his twelve tasks, the feat of obtaining some of these "golden apples of Hesperides." As you know, he succeeded, although the giant Atlas almost tricked him into holding up the sky. Later Perseus, who slew the Gorgon (from whose body sprang Pegasus, the winged horse), must have also visited the Isles of Hesperides. Atlas, fearful that Perseus was going to steal some of the "golden apples," tried to force Perseus to leave. Perseus held up the head of Gorgon and turned Atlas to stone—and the Atlas mountains in northwestern Africa hold up the sky today.

There is also the story about the Grecian maiden Atlanta, who was as fleet of foot as she was beautiful.

Now an art museum in Luxembourg, this orangerie was one of first to benefit from additional light of transparent ceiling.

The largest orangerie ever built was for Louis XIV at Versailles in 1682. The main gallery (center to right) is 500 feet long.

She had many suitors, but her conditions for marriage included beating her in a foot race. Suitors who did not were beheaded. One of the suitors, Hippomenes, obtained some of the golden apples and beat her in the foot race by rolling the apples at her feet as she passed him. She stopped to pick them up, and he kept running and won.

CITRUS AND RELIGION

Moses is thought to have referred to the citron when the Bible quotes him: "You shall take, on the first day, fruits of the tree *hadar*, or palm branches, boughs of the thickest trees, and willows that cross the length of rapid waters and rejoice before the Lord your God."

If the *hadar* is an ancient name for the citron, it explains the present day Feast of Tabernacles ritual. Practicing Jews appear once a year at their synagogue with myrtle, willow and palm boughs to which citrons are attached.

There is a variety called 'Etrog' that is commonly used but any citron meeting certain requirements is acceptable. For example, the fruit must come from an ungrafted tree; must be symmetrical without defects such as wounds or scars; and be used when relatively small.

For many Jews, the citron is the "apple" of the Garden of Eden. Others believe that a bite of the 'Etrog' citron will ensure a male child for a mother-to-be.

The most unusual citron is the *fingered* or 'Buddha's Hand' citron. It has been prized for centuries in Indochina, China and Japan. The fruit is more or less open at one end and divided into a number of finger-like sections.

The huge fruit of the *pummelo*, a close relative of the citron, is sacred to some Buddhist sects. Pummelo leaves floating in a child's bath water on Chinese New Year's eve purifies the soul of the child for the coming year. In areas east of Canton, newborn babies are traditionally bathed in water containing pummelo leaves. They cleanse the child of what we would call "original sin." Today, many Chinese use the pummelo fruit as a part of the centerpiece at their New Year's feast. The pummelo is native and grows to perfection in Tahiti, Taiwan, Java, Nagasaki Prefecture of Japan, Thailand, Vietnam and Indonesia. It has been introduced to both Hawaii and California.

CITRUS AND THE CONQUERORS
Citrus seeds were a part of the baggage bought by the conquerors of the New World. As the Spanish invaded Mexico, warriors began planting citrus. Bernal Diaz del Castillo, who, with Cortez, kept a record of his adventures, said they first landed at Cuba where Diaz picked up some orange seeds. From there, they left for what is now Vera Cruz. In Mexico, Diaz wrote, "There were so many mosquitoes near the river that ten of us went to sleep in a lofty Idol house, and close by that house I sowed the orange seeds." Diaz planted the seeds near what is present day San Anton. By the 1700's, Spanish missionaries were planting citrus in California and northern Sonora, Mexico.

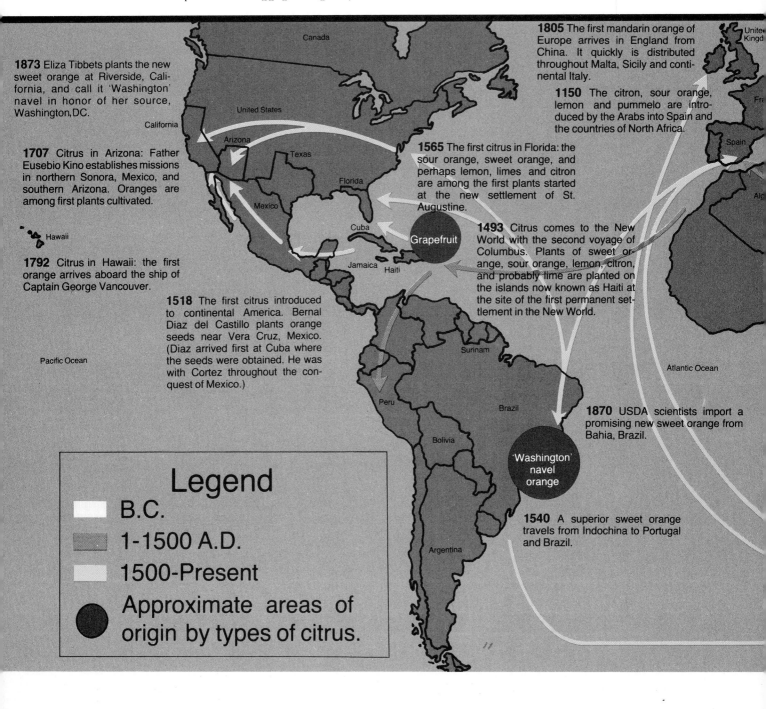

1873 Eliza Tibbets plants the new sweet orange at Riverside, California, and call it 'Washington' navel in honor of her source, Washington, DC.

1707 Citrus in Arizona: Father Eusebio Kino establishes missions in northern Sonora, Mexico, and southern Arizona. Oranges are among first plants cultivated.

1792 Citrus in Hawaii: the first orange arrives aboard the ship of Captain George Vancouver.

1518 The first citrus introduced to continental America. Bernal Diaz del Castillo plants orange seeds near Vera Cruz, Mexico. (Diaz arrived first at Cuba where the seeds were obtained. He was with Cortez throughout the conquest of Mexico.)

1805 The first mandarin orange of Europe arrives in England from China. It quickly is distributed throughout Malta, Sicily and continental Italy.

1150 The citron, sour orange, lemon and pummelo are introduced by the Arabs into Spain and the countries of North Africa.

1565 The first citrus in Florida: the sour orange, sweet orange, and perhaps lemon, limes and citron are among the first plants started at the new settlement of St. Augustine.

1493 Citrus comes to the New World with the second voyage of Columbus. Plants of sweet orange, sour orange, lemon, citron, and probably lime are planted on the islands now known as Haiti at the site of the first permanent settlement in the New World.

1870 USDA scientists import a promising new sweet orange from Bahia, Brazil.

1540 A superior sweet orange travels from Indochina to Portugal and Brazil.

Grapefruit

'Washington' navel orange

Legend
B.C.
1-1500 A.D.
1500-Present
Approximate areas of origin by types of citrus.

Citrus Travels

The evocative names of many of the citrus varieties today are the record of their travels. In the map below we have traced some citrus migration. There is the 'Seville' sour orange, named for the city of southern Spain; or the 'Persian' and 'Tahitian' limes. The 'Jaffa' ('Shamouti') orange is named for a city near Palestine. The well known 'Valencia' orange is named for the region of eastern Spain where it probably originated, likewise the 'Lisbon' lemon.

Citrus names can mean fragrance: France knows the 'Bouquet de Fleurs'. It is the especially fragrant flower of a sour orange used for perfumes.

Citrus names can mean color: from Spain is the 'Sanguinelli' or blood orange.

The citrus heritage is a rich one, more so than any comparable fruit or food plant. Like a pet, they went everywhere men did as the world was explored. Now modern home gardeners are rediscovering the potential of citrus. Many will adopt a tree or two.

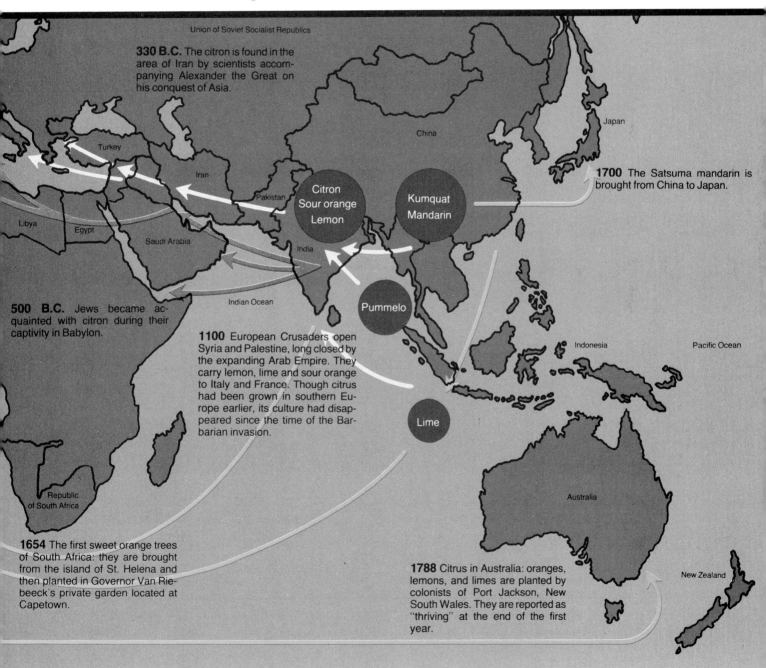

330 B.C. The citron is found in the area of Iran by scientists accompanying Alexander the Great on his conquest of Asia.

1700 The Satsuma mandarin is brought from China to Japan.

500 B.C. Jews became acquainted with citron during their captivity in Babylon.

1100 European Crusaders open Syria and Palestine, long closed by the expanding Arab Empire. They carry lemon, lime and sour orange to Italy and France. Though citrus had been grown in southern Europe earlier, its culture had disappeared since the time of the Barbarian invasion.

1654 The first sweet orange trees of South Africa: they are brought from the island of St. Helena and then planted in Governor Van Riebeeck's private garden located at Capetown.

1788 Citrus in Australia: oranges, lemons, and limes are planted by colonists of Port Jackson, New South Wales. They are reported as "thriving" at the end of the first year.

Union of Soviet Socialist Republics

Turkey

Iran

Pakistan

Citron Sour orange Lemon

Kumquat Mandarin

China

Japan

Libya

Egypt

Saudi Arabia

India

Indian Ocean

Pummelo

Lime

Indonesia

Pacific Ocean

Republic of South Africa

Australia

New Zealand

Citrus: Drenched in History

2201 B.C. "The baskets were filled with woven ornamental silk. The bundles contained small oranges and pummeloes. . . ." From the ancient book, Yu Kung, meaning Tribute of Yu, written about Chinese emporer Ta Yu.

586-539 B.C. During their captivity in Babylon, the Jews became acquainted with the citron which had been cultivated in Mesopotamia before then.

310 B.C. The citron is mentioned in the writings of the Greek historian Theophrastus. It is the first citrus of any kind written about by a European. He wrote: "Thus one sees in Media and Persia among many other productions the tree called *Persian* or *Median apple* . . . Its fruit is not edible but it has an exquisite odor, as also have the leaves which are used as a protection from moths in clothing. A decoction of the pulp of this fruit is thought to be an antidote to poison, and will also sweeten the breath. . . The citron bears fruit continuously. While some fruit is falling with ripeness other fruit is but just starting. . . Fruit is given only by the flowers which have in the middle a sort of straight spindle; those which do not have this fall off, producing nothing."

70-19 B.C. The Latin poet Virgil includes mention of the citron in his writings. He uses the name Median apple.

A.D.

60-79 Dioscorides of Cilicia, the area along the coast of southern Turkey, writes in his *Materia Medica* of the citron as very well-known in his area. He uses the names Median and Persian apple as well as "cedromela."

77 Pliny, the Roman naturalist and writer, talks of the citron as a medicine, poison antidote, perfume and moth repellent. He is the first to use the word "citrus." Other names he cites are *malus medica* and *malus assyria*.

50-150 Trade routes to and from Rome are well established. Gourmets in Rome occasionally have citrons, lemons and oranges.

300 Italian gardeners overcome difficulties in getting fruit from trees grown in that country. Apparently citrons, lemons and oranges were grown. Mosaics of all three decorate the mausoleum of Lady Constantia, daughter of Constantine the Great.

568 The Lombard invasion breaks the link between Italy and the Byzantine Empire. The luxurious gardens of the rich are destroyed and with them the delicate citron, lemon and orange trees. Without some shelter and special care, the trees eventually die out. Some citrons and oranges survive in the wild where climate was most favorable (Sicily, Sardinia and the region of Naples).

570-900 The Roman Empire continues to disintegrate as the Arabian Empire expands. Moslem culture makes significant contributions to citrus distribution by introducing the plants into countries they control: Persia, Iraq, Syria, Palestine and Egypt.

1003 A prince of Salerno, an Italian seaport near Naples, sends citron fruit to Norman Lords in appreciation of their protection from the Saracens.

1100 The lemon is introduced into the area known as the Kwangtung Province of China.

Historical citron variety known as 'Buddha's Hand'.

1150 The citron, sour orange, lemon and pummelo are introduced by the Arabs into Spain.

1175 The first reference to the lemon in Chinese literature: Fan Ch'eng-ta writes that the "li-mung," or lemon, is "the size of a large plum . . . (and) resembles a small orange and is exceedingly sour to the taste."

The Mongolians are believed to be the first to use lemonade.

1178 Han Yen-chih writes this year a *Monograph on the Oranges of Wen-chou, Chekiang.* In it he names and describes 27 varieties of sweet, sour and mandarin oranges. He also describes citrons, kumquats and the trifoliate orange and discusses nursery methods, grove management and diseases.

1250 The first use of the term "arangus," meaning oranges, is by Albert Magnus in his *Libre de vegetabilibus* with his description of the sour orange. ". . . The fruit is short and round, the tree is larger and more cold resistant than the citron, and the leaves appearing to be divided into two, the largest toward the end standing above the smaller one . . ."

1250 The first known mention of the lime in literature is made by the Arab writer, Abd-Allatif. Called the "balm lemon," it is described as having smooth skin and being about the size of a pigeon egg.

1299 History of Mongolia includes a reference to lemonade being popular at this time.

1400 Orangeries, greenhouses especially for citrus, become popular in Europe. Earlier accounts mention several techniques used to protect tender citrus. But by this time whole buildings with artificial heat and climate were built especially for citrus and oranges. Though not the first or only use of the greenhouse principal, citrus culture prompted the first widespread use.

1493 Citrus comes to the New World with the arrival of Columbus at Haiti. Documentation is scanty, but does state: ". . . During this time, with great haste, he (Columbus) provided himself with some cattle, which he and those that came with him brought . . . They brought hens and also grains, and seeds of oranges, lemons, and citron and all kinds of melons. . ."

1518 First citrus of continental America. Bernal Diaz del Castillo plants orange seeds near Vera Cruz, Mexico.

1523 Monk Leandro Alberti reports immense plantations of oranges, lemons and citrons are thriving in Sicily, Calabria (southwestern Italy opposite Sicily), and along the river Salo in Liguria (northwestern coast of Italy). He also noted that many varieties were cultivated and most were sweet.

1525 Lemons, limes and citrons are so abundant on the island of Haiti that "they are now past counting," writes Ovieds y Valdez, a naturalist traveling to Santo Domingo, Haiti that year.

1565 Citrus planted in Florida at settlement of St. Augustine.

1575 Oranges planted on Parris Island, South Carolina.

1590 Orange conserve, or marmalade, is popular with the inhabitants of the West Indies. "The best I have found anywhere," according to Acosta, a traveler there that year.

1646 The Italian botanist Ferrari describes the Portugal orange. This orange, originally from China, was the first significantly superior sweet orange of Europe. The Portugal orange stimulated the European citrus industry as dramatically as the 'Washington' navel did in California some 250 years later.

1654 Sweet orange trees are planted in the private garden of the governor of South Africa.

1677 The first known use of the word "lime" in literature: Sir Thomas Herbet wrote of finding "oranges, lemons and limes" on the island of Mohelia off the coast of Mozambique during a voyage there this year.

1707 Father Eusebio Kino plants citrus in northern Mexico and Arizona.

1750 The grapefruit, probably a mutation of the pummelo, is first described in literature by Griffity Hughes reporting on his travels to Barbados. He calls it the "forbidden fruit."

1788 Oranges, lemons and limes are planted in Australia by original settlers.

1769 Father Junipero Serra plants citrus at Mission San Diego.

1792 Captain George Vancouver arrives at the island of Hawaii with orange seedlings aboard ship.

1805 Europeans are excited by the "mandarian" orange, newly arrived from China. Soon mandarins are growing in Malta, Sicily and Italy.

1821 Florida becomes a part of the United States. The fledgling citrus industry there is stimulated by the subsequent opening of new markets in the north. St. Augustine groves annually produce 2 to 2½ million oranges. They are shipped to Charleston, Baltimore, New York and Boston and are sold for one to three dollars per hundred.

1830-1870 Wild orange groves in Florida (many several acres in extent) are improved by grafting. The practice is so successful and profitable that most are converted.

1834 First California citrus planted outside a mission: Jean Louis Vigens procures from Mission San Gabriel 35 large seedling sweet oranges which he plants at his house on Aliso Street in Los Angeles.

1835 Don Phillippe, a Spanish nobleman, plants the first grapefruit in Florida.

1856 Judge Joseph Lewis plants three sweet oranges in the Oroville vicinity, north of Sacramento. The first citrus of northern California, one still survives and is called "The Mother Orange." The oldest orange tree in the state, it was transplanted to its present site overlooking the Oroville Dam in 1965.

1862 Two orange trees are planted in Frasier Valley, east of Porterville, Tulare County, California. The area is now one of the most important citrus growing belts of the state.

1867 The United States Department of Agriculture reports 17,000 orange trees and 3,700 lemon trees in California. Most are in the Los Angeles area.

1869 Cottony-cushion scale is unwittingly imported into California along with some Australian acacia trees. Within a decade the pest threatens to destroy all citrus plantings in the state.

1870 Small or large citrus orchards are thriving in all the principal citrus growing regions of California and the commercial industry is well established.

One of the early rail shipments of citrus from west to east.

1841 The first commercial citrus grove is planted in California. The site chosen by William Wolfskill is to become the Arcade passenger station of the Southern Pacific Railroad in downtown Los Angeles.

1849 The California gold rush and the rapid population growth that follows is the real birth of the commercial citrus industry in the Golden State. Fruit is shipped from Los Angeles to San Francisco, and then up the Sacramento, American and Feather Rivers to points near the gold mines.

1870-1875 Transcontinental railroads are completed. Eastern markets are potentially open to California citrus.

1874 The famous orange to become known as the 'Washington' navel is planted in Riverside, California. Eliza Tibbets plants them at her doorstep.

1877 William Wolfskill oranges from Los Angeles reach St. Louis after a month in rail transit. The shipment marks the first eastern sale of California citrus.

Plaque at site of parent 'Washington' navel in Riverside.

1880 Citrus of all kinds are flourishing in Riverside County, California. Northern groves are not yet significant and Los Angeles County produces slightly more than Riverside.

1880 The first shipment of Florida grapefruit is made to Philadelphia and New York. For many years previous, the grapefruit was considered a novelty fruit and most were left to rot on the ground.

1886 The first citrus field laboratory is established by D. W. Coquillet, an entomologist, on the Wolfskill ranch, Los Angeles, California.

1889 Control of California's cottony-cushion scale is achieved by biological means. Albert Koebele of the USDA returns from Australia with the Vedalia or Australian lady beetle, a natural predator of the pest. Vedalia so successfully eradicates this serious pest that it is acclaimed as the most spectacular demonstration of biological control to this day.

1892 Five carloads of California citrus are shipped to England by the California Fruit Transportation Company. They went by rail to New York, then took a 14-day steamer voyage to Liverpool and London. Great crowds attended the first sale of California citrus in London. Queen Victoria sampled some California oranges from the shipment.

1892 The first cooperative citrus exchange, the Pachappa Orange Growers Association, is established in California.

1892 The USDA establishes the Subtropical Laboratory at Eustis, Florida. Two researchers, Webber and Swingle, make the first study of citrus diseases—blight, dieback, foot rot and scab.

1893 The Southern California Fruit Exchange is established to promote and cooperatively market the increasing citrus production.

1900 The first commercial citrus orchards are planted in Arizona.

1907 The Citrus Experiment Station at Riverside, California is established. Much of the original research regarding citrus culture and varieties begins now. Still in operation, it is known today as the Citrus Research Center and Agricultural Experiment Station.

1910 The first commercial plantings of citrus are made in the lower Rio Grande Valley of Texas.

1916 Citrus culture in Texas spread northward into the Houston and Beaumont areas only to be destroyed by severe frost this year. Commercial culture is restricted to the lower Rio Grande Valley.

1948 Frozen and hot-packed citrus juice is marketed across the U.S.

1949 Four Winds True Dwarf Nursery sells the first "true dwarf" trees. It is the culmination of many years experiments and the beginning of a new industry of particular benefit to the home gardener.

1952 The Southern California Fruit Exchange changes name to Sunkist Growers.

1962 The U.S. now grows about 34 percent of the world's oranges, 85 percent of the grapefruit, and 45 percent of its lemons. Arizona and California are the major producers of fresh fruit; Florida's citrus industry is geared to juice production.

TODAY—A wide range of citrus varieties are now available for the home gardener's own experimentation and pleasure. The commercial industry grows by improved production techniques, improved post-harvest handling and a widening range of citrus by-products.

Citrus In the West

If one statement characterizes the movement of citrus into Arizona and California, it would be "citrus followed the cross." By the early 1700's, Spanish missionaries and explorers had firmly established themselves in northern Mexico. They had many familiar fruits from their home country: olives, pomegranates, figs, mulberries and many others, including citrus. The missionaries, anxious to spread the word of Christianity, began to look toward more northern areas reportedly occupied by savage Indians.

The first settlements in southern Arizona were made in 1683 by Father Eusebio Kino.

When citrus was introduced to California by Franciscan Father Junipero Serra in 1769 at San Diego, a pattern for development was set. Next to setting up an altar and barracks, planting a garden and securing food was the most important task. Lack of water was a problem but the missionaries were good gardeners and the plants, especially citrus, flourished in the California climate.

As the missions spread, so did citrus. The fathers set up nurseries to grow plants to supply expeditions farther north. The fruit played a role in the conversion of the Indians. Orange juice and lemonade helped sermons make converts. The priests found that the way to save souls was through the belly as well as the spirit.

The gardens around the missons were cherished possessions and the priests were often reluctant to give seeds or plants of their citrus to settlers. Many of the gardens were fenced by heavily thorned cactus.

Gradually attitudes changed and by the 1830's small private groves began to spring up throughout southern California. The most important belonged to William Wolfskill.

FIRST CALIFORNIA ORANGE GROVE

In 1841, William Wolfskill, a Kentucky trapper of German ancestry, planted his first orange tree on the site which was to become the Arcade passenger station of Southern Pacific Railroad in Los Angeles. Despite ridicule from neighbors who could not imagine selling oranges for profit, Wolfskill kept on planting and the idea caught on. He gradually increased his plantings until he had more than 16,000 trees on 28 acres. His last crop sold on the trees for a whopping $25,000.

There were a number of reasons for his success. The California Gold Rush was in full swing. Famous Wolfskill oranges were shipped by rail to San Francisco, then by ship up the Sacramento, American and Feather Rivers right to the miner's camps.

The completion of the transcontinental railroad also contributed heavily to Wolfskill's success. In 1877, Wolfskill sent a carload of oranges to St. Louis. Even though they took a month to get there, they arrived in good condition. Thus began the eastern identification of the California climate with citrus.

THE WASHINGTON NAVEL

In 1873, William Saunders, Superintendent of Gardens and Grounds for the United States Department of Agriculture, gave Eliza Tibbets of Riverside, California, three branches of a new orange variety that he obtained from Bahia, Brazil. It was first grown in Washington, D.C., so was named the 'Washington' navel. Quality was superior to any other orange grown in California.

The three trees started fruiting in 1878. The response to these oranges was so great and the demand

The three original 'Washington' navel sweet oranges were planted by Eliza Tibbets, below, at her Riverside, California home in 1873.

so high that Eliza could sell propagating material for up to $5.00 a piece. Today, one of the three original trees is still alive and bearing fruit in Riverside.

CITRUS IN NORTHERN CALIFORNIA

During this same period of the mid-1800's, citrus trees were also being planted in northern California. The most famous planting was that of Judge Joseph Lewis who planted three sweet orange seedlings at Bidwell's Bar outside of Oroville in 1856. One of these trees is still producing fruit although it was moved in 1965 to a site above the Oroville Dam. It is known as the "Mother Orange" and is evidently the oldest living orange tree in California.

By 1870 there were orange trees bearing fruit in the Sacramento, Marysville and Porterville areas and even as far north as Red Bluff.

GROWING PAINS

The period from the mid-1850's to the early 1900's was a time of failure as well as success for early California citrus growers. With the lure of gold and warm sunny days, easterners migrated to California. Between 1880 and 1890, the population of California grew by 343,436 people to a total of

1,208,130. Many of these people risked everything to come west and make a fortune in either gold or citrus. In southern California thousands of acres of citrus were planted by these settlers. Eventually these groves would be worth fortunes, but the cost of the land, planting and maintenance for several years before the trees produced fruit was too much for some "boom-or-bust" newcomers. Many went broke, abandoned their groves and returned east.

Those that stuck it out learned from their experience. They noted certain areas were better for particular varieties. Navels did especially well in the Riverside area. Parts of Orange County were ideal for Valencias. Growers in coastal regions of Ventura, San Diego and Santa Barbara counties found they could harvest lemons several times a year.

CITRUS COOPERATIVES

Eventually the vast acreages of citrus that were planted during California's period of rapid growth came into production, presenting growers with a new problem: Too much citrus. While the eastern market was eager for citrus but could not obtain it, growers found all their fruit reaching the market at the same time and much was wasted.

Transplanting one of the original 'Washington' navels was cause for ceremony in 1903. It survived only a few years in front of what is now the Mission Inn in Riverside.

Another of the original trees was transplanted at the same time to this location at the head of Magnolia Street in Riverside where it survives to this day.

Cooperative marketing was the solution. Growers in the Claremont and Riverside areas joined to share expenses of packing and marketing. Together they "weighed, culled, graded, packed, sold, shipped or otherwise disposed of" the citrus. Each grower was paid according to his production. By controlling the shipments, the cooperative oranges, which bore a special label, brought in high prices while some independent growers were going broke. Needless to say, the idea caught on quickly.

On August 29, 1893, several of the cooperatives and independent growers joined together under the name Southern California Fruit Exchange. By 1894, the Exchange claimed to handle 80 percent of the oranges grown in southern California.

The Exchange struggled for several years until experience was gained in marketing fresh fruit. By 1905, its name was changed to the California Fruit Exchange to encompass citrus growers in northern California. The cooperative growers then set out on aggressive marketing and advertising campaigns that would forever change the face of the fresh fruit industry.

Aimed at midwesterners, the California Fruit Exchange advertised fresh citrus as a "warm ray of California sunshine." They labeled their citrus "Sunkist" and easterners consumed oranges at an amazing rate. Citrus labeled Sunkist soon came to represent the "cream of the crop."

Actually, the name Sunkist means more than just effective marketing. The California Fruit Exchange pioneered shipping, packing and advertising techniques. They helped establish fresh orange juice as America's breakfast drink, and promoted a wide array of citrus products. Their growth is an important part of California's agricultural history.

DWARF CITRUS—A RECENT DEVELOPMENT

In 1946, a curious and soon-to-be nurseryman named Floyd Dillon decided there was a need for a California version of the dwarf apple and pear. He chose citrus as the ideal candidate. As he later wrote in 1959: "What better specifications could be written for the ideal patio plant than to take everything offered by the standard commercial citrus and dwarf it to patio size, capable of being grown in a box or tub, raised bed or border, either espaliered or in its natural form?

"The orchard citrus of most species and cultivars grows in a dense, globular form. Its evergreen foliage

A Pomona, California, citrus packing house in 1907. By this time marketing cooperatives were well established and fruits were shipped by rail and steamer to distant locations.

is clean and attractive, with almost a continuous showing of fragrant flowers or waxy colorful fruits. It is decorative, interesting and productive.

"But at maturity, the standard citrus is 20 feet or more in width and occupies more than 400 square feet of garden space. It comes into bearing slowly. Only a few forms are available at retail nurseries.

"If the virtues of citrus could be had in an eight-foot tree, capable of producing fruit two years after planting, if numerous cultivars and novelties were available, if all cultivars could be uniform in size—then we would have the ideal patio tree."

Floyd soon learned that researchers at UCLA, the University of California Citrus Research Station at Riverside, and the USDA Date and Citrus Experiment Station in Indio, California were already experimenting with new varieties and different rootstock—*scion* combinations. A scion is a twig or branch of superior eating variety grafted or budded to roots of another variety to improve disease resistance, production, control tree size or modify other characteristics.

With the help of well known citrus researchers such as William Bitters, author of *Dwarfing Citrus Rootstocks* in 1949, and many others, Floyd chose rootstock/scion combinations and propagation techniques that gave him the desired results.

With a wealth of new information and several years of experimentation, promotion and family help, Floyd founded Four Winds True Dwarf Nursery which now makes its home in Mission San Jose, Fremont, California.

Today, dwarf citrus is a common sight in nurseries throughout the West and is featured in mail order catalogs nationwide.

The adaptability of dwarf citrus to container growing has made them increasingly popular outside of traditional citrus growing areas. They are most easily moved indoors for winter protection and outdoors in the warm seasons.

In addition, dwarf citrus has become important to gardeners whose gardening space is limited to patios, balconies and small porches.

It seems Floyd Dillon and his dwarf citrus nursery were well ahead of their time. As interest of non-commercial gardeners grows, more varieties of citrus are becoming available on dwarf rootstocks. Blood oranges, for instance, are rarely available as fresh fruit, but are becoming a home garden favorite.

THE CITRUS LEGEND CONTINUES

Citrus, once eaten only by kings or the wealthy, is now available to everyone, Best known for its vitamin C content, citrus is generally perceived as a kind of health food. Citrus versatility is impressive—for example, more lemons are processed for their flavors than are sold for fresh use.

The fifth volume of the respected *Citrus Industry*, recently published, makes available the most factual survey of scientific citrus research. No other tree fruit has been studied and reported as much as citrus.

Citrus continues to be the singular interest of several federal, state and commercial interests. For instance, in California there is the highly regarded University of California Agricultural Center at Riverside, the Lindcove Field Station at Exeter, the Kearney Horticultural Field Station at Parlier, the USDA Citrus Research Station at Indio, the Sunkist organization in Ontario.

In Arizona there is the Yuma Field Station and the Tempe Field Station. In Florida it is the Lake Alfred Citrus Experiment Station, the Orlando Research Station—head office and chief facility for USDA citrus research, and the extensive citrus products industry.

Through hundreds of years citrus has received more attention than virtually any other fruit plant, and it continues to do so. Any plant receiving this much attention is bound to know many changes. Climate adaptability has broadened and varieties have improved. It's likely this pattern will continue. There is talk today of larger fruits being introduced, perhaps uniformly seedless.

Home gardeners can look forward to the possibility of smaller size dwarfs. Rootstocks are being worked with now that will confine tree growth to as little as three feet compared to the typical eight feet of today's dwarf.

An increasing number of new varieties will become available, and citrus of all kinds will assume a larger role in landscape plantings.

Although the pressures of urbanization and catastrophic disease problems have forced relocation of many groves, the citrus industry in the United States and the rest of the world is thriving.

THE CITRUS INDUSTRY TODAY

Both Arizona and Texas are now important commercial citrus producing states, even though planting was initiated years later than California. The United States is the leading citrus producing country in the world. The Food and Agriculture Organization of the United Nations estimates over 15 million metric tons will be produced in the U.S. in 1985. This compares to the projected total world citrus production of over 63 million metric tons for that year.

From Tree to Market

Citrus gardener or not, you are likely to make at least occasional use of the market supply of fresh oranges, lemons, limes or juice concentrates. It's a remarkable service the industry provides citrus lovers. Here is how they do it.

HAND PICKED

Because of equipment costs and difficulties involved in machine harvesting, practically all citrus is picked by hand. The harvest remains essentially a "bag and ladder" operation. A stout cloth bag that opens at the bottom is carried by a strap across the picker's shoulder. With the aid of a strong, lightweight, "straight" ladder, a special hand clipper and soft cotton gloves, trees are picked from the top down. "Skirt" picking is done from the ground. When the bag is full, it is carried to a nearby bin. In order to prevent fruit injury it is separated from the stem at the "button" where it is attached. A specially designed snub-nosed clipper is used. Fruits are harvested only when the fruit and foliage are dry.

Moisture on the fruit makes the rind turgid. A turgid rind is subject to damage due to the release of

Loading bins of citrus—the beginning of their travels from orchard to you. From here they are sent to either packing house or processing plant and from there to market.

Fresh fruit "culls" are transported by truck to a processing plant. There they are utilized in one of many special citrus products.

Every ten minutes around the clock and every day of the year, a 25-ton load of citrus is delivered to a Sunkist processing facility.

rind oil when pressure is applied. Any injury gives admittance to disease organisms which cause fruit to rot. An unbroken, healthy skin is protection against almost all decay. When the rind is abraded or punctured, the fruit becomes infected by ubiquitous disease spores. Rough handling due to careless dropping or overfilling the bins adds to the disease problem.

Bins holding from 900 to 1,000 pounds of fruit are placed in convenient locations in orchard picking "drives." Machines pick them up and load them onto trucks when full. The fruit is delivered directly to the packing house where it moves by forklift to storage rooms.

THE PACKING HOUSE

How citrus is handled in the packing house depends on the variety, stage of maturity and condition on arrival. Because lemons are usually picked green, they must undergo a relatively long curing period in storage. When the fruit is "prime," it often goes rapidly through the packing procedure and is shipped directly to the market. Mandarins are more delicate, particularly when harvested early, and should be handled with the care given to fresh deciduous fruit. Preferably, they are "dry brushed," sized, color-sorted and sent to market with comparatively little storage and handling.

The packing house manager judges the condition of each lot of fruit and gets it ready for shipment. He assigns similar lots to the most suitable destination and sees to it that the fruit moves under the best possible conditions.

To provide a view of packing house operations, let us look at how navel oranges are handled. Much of the fresh citrus fruit grown in the West is handled in a similar way.

Early season navel oranges usually command a premium price. As a result there is a mad scramble to pick when the market is "hot" and to move a sizeable portion of the crop before damaging frost occurs. To avoid glutting the market, particularly with immature fruit, quality standards and a system of controlled marketing (pro-rate) have been adopted by the citrus

To this Ontario, California, plant come navel and 'Valencia' oranges, white and ruby grapefruit and 11 varieties of tangerines. All will be converted into juice concentrates and peel products. In the photo opposite, the fruit is moving by conveyor into the plant. Above, metal drums of processed juice concentrate move out of plant into cold storage.

industries of California and Arizona.

Quality standards include a minimum soluble solids (sugar) to acid ratio (8:1 is minimum sugar-to-acid ratio), acceptable color and size, as well as freedom from serious frost injury and other external and internal defects. Orderly marketing is assured by a federally administered grower-handler Marketing Order. Policy is set by the pro-rate committee, which meets once a week during the shipping season. It determines how much fruit to release to the market in relation to supply and demand. Its overall objective is to move the crop in an orderly fashion throughout the entire period of harvest. This provides the consumer with a steady supply of oranges—presumably at fair market prices.

GRADING

When the fruit handler decides the oranges are suitable for packing and the market is ready, the oranges are dumped into a hopper which feeds them onto a receiving belt. A small crew, working in a darkened area with the aid of fluorescent lighting,

detect and discard rots and splits. The oranges then drop into a warm, soapy wash and are thoroughly washed and brushed. They are then picked up by another conveyor and dried by blasts of warm air. "Graders" are stationed at intervals that permit a thorough inspection of the oranges. Bruised, cut, misshapened, scarred and other undesirable fruit ("culls") are separated and placed on the proper conveyances. Sound oranges are diverted to two or more grade designations, after which the oranges go through an elaborate sizing machine.

Oranges are sized on the basis of number of fruit per carton. Each carton will contain oranges of uniform size with a total weight of approximately 36 pounds. One thousand cartons fill a standard railroad box car.

After sizing, the oranges move onto a series of slanting tables where workers are stationed. The carton is filled, layer by layer, and is placed on a belt which conveys it to a machine that automatically seals the carton. Finally, the carton is conveyed to a receiving area where it is stacked on a pallet with several others

Upon entering the processing facility, the fruit is thoroughly cleaned. Such efficient use is made of the fruit that by the end of the process every bit of juice, pulp, oil, seeds and peel is used for many special products.

After washing, the fruit goes through a final inspection to eliminate those not consistent with quality standards.

containing oranges of the same size and grade. A forklift is used to move the pallets into a cold storage room. There the cartons are held under carefully controlled temperature, humidity and aeration until placed in refrigerated trucks or box cars.

Whether by truck or rail, the oranges move quickly to their destination—wholesale produce market or supermarket warehouse. Within a day or two, oranges are placed on the fruit stand where you may take your pick of one of nature's finest products, ready to eat in its own germ-free, biodegradeable wrapper!

ORANGE JUICE

Oranges intended for juice and other products are processed at a few large plants in California and Florida. Frozen citrus concentrate is by far the most important manufactured item. Fruit that cannot meet quality standards for the fresh market or fruit that is particularly good for juicing is delivered by huge semi-trailers from orchards as far as 250 miles from the plant. On arrival, the fruit is dumped into huge bins that provide a reservoir so that plant operation may continue with an uninterrupted fruit supply.

Unlike citrus destined for the fresh market, the grower is paid on "pounds-solids" basis, and not fruit-quality basis such as that of taste, aroma and eye appeal. The fruits culled from juice lots are "splits" and "rots." Sound fruit is washed and scrubbed before reaming or crushing. During the extraction operation, the juice is separated from the pulp and each is diverted into holding tanks.

After the juice is concentrated by evaporation, details in the further preparation of frozen concentrate vary from plant to plant. Because of its generally dependable good quality and wholesomeness, and the ease with which it is reconstituted for home use, frozen concentrate has wide consumer acceptance. Although much citrus is marketed in one juice form or another, the major use for oranges is in frozen concentrate.

It is indeed a blessing that there are so many ways citrus can be used. And it is most fortunate for the lover of citrus juice that a year-round supply of frozen concentrate is available even at times when adverse weather destroys the crop.

Packaged juices may be juice of a single variety or a blend of several. Technicians follow every batch through every step, testing for purity and conformity to standards.

High speed production lines for glass products fill glass bottles and jugs ranging in size from a few ounces to a gallon. Some by-products are shipped long distance in 10,000-gallon railroad tank cars.

Fresh fruit, juice and many other citrus products reach their destination at the supermarket.

SUNNY HEIG

GROWN ON THE SUNNY HEIGHTS

PACKED BY

REDLANDS CO-OPERATIVE

REDLANDS, CALIFOR

Sunkist

HTS

EDLANDS

JIT ASSN.

2 Citrus Climates

Where you live—Ventura, El Paso, Tucson or New Jersey—will largely determine the citrus varieties you can grow and how they will behave. Variety recommendations, ripening times and other pertinent information is included in the chapter "Which Varieties Will You Choose?" But consider the information here about cold hardiness and microclimates. It tells how to "stretch" your climate and grow varieties not typical in your area.

The effects of climate on citrus may be subtle or dramatic. To illustrate them, we will take a latitudinal trip. Beginning in the tropical climate of the small South American country of Surinam located northeast of Brazil, we'll travel north to the two major commercial citrus areas of the United States, and compare the semitropical climate of Florida to the subtropical climate of California. We will look at the effects of climate in the three major growing areas of California: coastal, inland and desert.

CITRUS IN TROPICAL CLIMATES

Tropical climates are generally defined as areas between 15° south latitude and 15° north latitude; that is, 15° north or south of the equator.

Tropical climates are characterized by high heat and humidity. Seasons as we know them in the United States are relatively nonexistent. There is very little fluctuation between day and night temperatures. Rain falls evenly throughout the year. Dry spells are short.

The small country of Surinam is north of Brazil and lies between 4° and 6° latitude, north of the equator. It has a climate similar to that described above.

The first thing you might notice about the citrus in Surinam is that the fruit appears green when marketed, yet internal color is the same as the fruit you buy. The bright coloration on the outside of most

This old crate label extols the joys of the California climate. Citrus crate label art reached its peak in 1900 to 1930. It died when the wooden crate was replaced by the cardboard carton in the 1940's.

citrus depends largely on daily temperature changes. Cool nights are necessary for the skin to become brightly colored. Surinam nights are warm, so the rind is still green when the fruit is ready to eat.

The effects of heat (or the lack of it) on citrus are dramatic and varied. Most of these effects are compounded in Surinam and present some unusual problems for citrus growers there.

In tropical climates, citrus trees grow rapidly, bloom often and set fruit the year-round. A prolific bloom usually comes with heavy rains after a dry spell. This growth habit is problematical because there are many crops of fruit on the trees at various stages of maturity, and all are some shade of green. The constant heat causes the fruit to age rapidly on the tree so there is no holding period after maturity. When the fruit is ripe, it must be picked.

So which fruit do you pick? You probably won't be able to tell. Some you choose will be underripe, some overripe. That's why most of the citrus grown in Surinam is sold only locally, and quality is unpredictable.

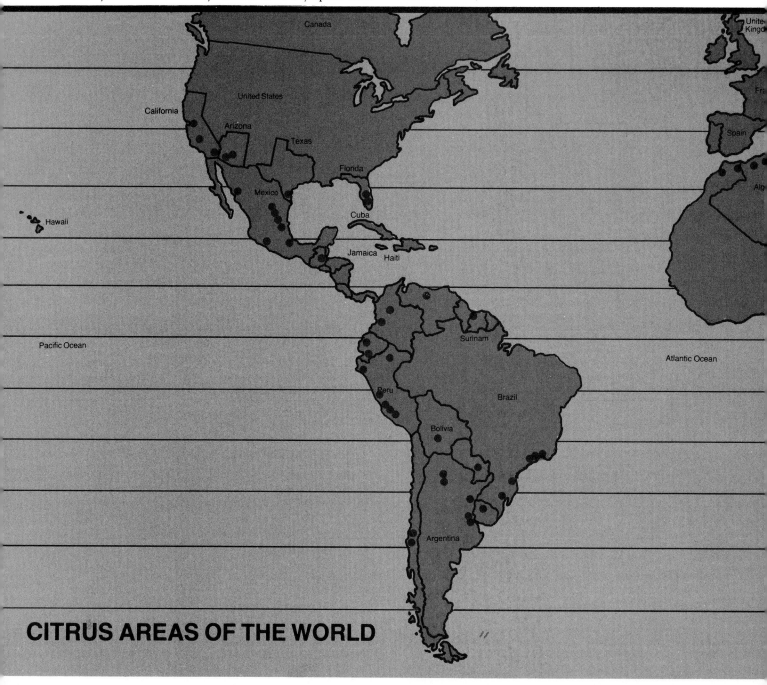

CITRUS AREAS OF THE WORLD

CLIMATES OF CALIFORNIA AND FLORIDA

California and Florida, between 38° and 28° north latitude respectively, are the most important citrus growing regions of the United States. Florida produces about three times as much citrus as California and is the most productive citrus area in the world. Its climate is *semitropical*, intermediate between Surinam's tropical climate and *subtropical* regions such as California. Summers are warm and humid. Winters are cool in most of Florida with some frost expected every year.

Heat and humidity are the main climatic factors that distinguish differences between Florida and California citrus. California has wider daily temperature fluctuations, so fruit is generally brighter in color. Daily temperature range also affects fruit flavor. Wide differences between day and night temperatures promote both sugar and acid formation. In tropical or semi-tropical climates, the acid flavor of citrus is not as pronounced as it is in subtropical citrus. The fruit tastes sweeter. Citrus flavor is judged by the ratio of sugar to acid. Sweetness alone is not a determinant of

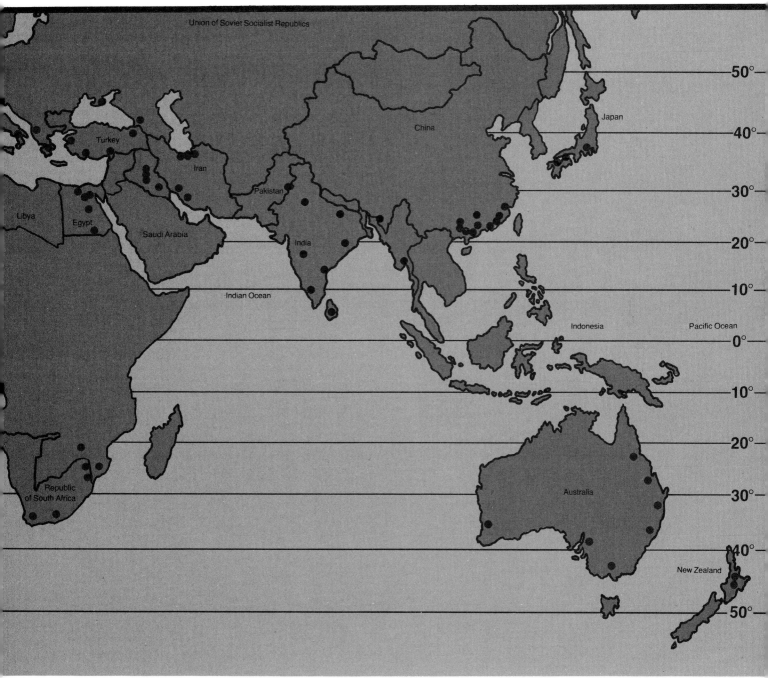

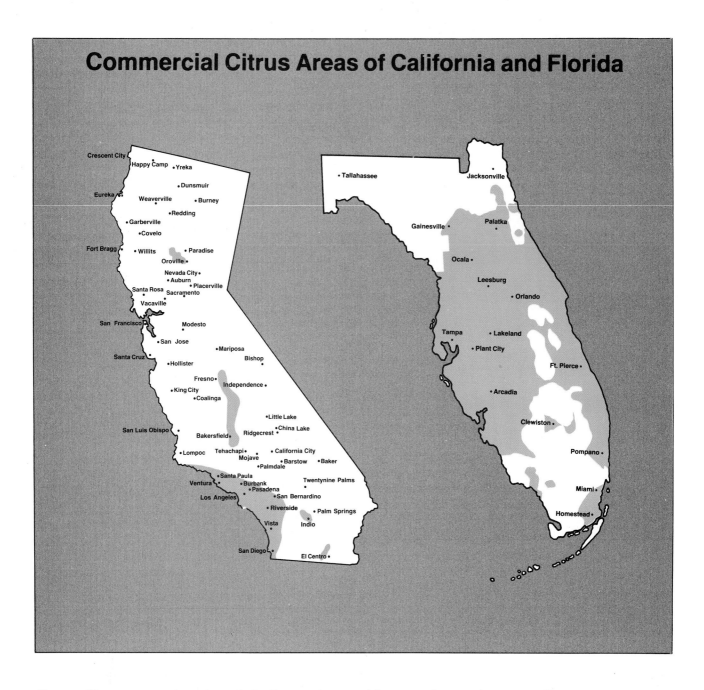

Commercial Citrus Areas of California and Florida

Crescent City
Happy Camp • Yreka
• Dunsmuir
Eureka • Weaverville • Burney
• Redding
• Garberville
• Covelo
Fort Bragg • • Paradise
• Willits
Oroville •
Nevada City •
• Auburn
Santa Rosa • • Placerville
Sacramento
Vacaville •
San Francisco • Modesto
• San Jose
Santa Cruz • • Mariposa
• Hollister Bishop •
Fresno •
• King City Independence •
• Coalinga
• Little Lake
San Luis Obispo • • China Lake
Bakersfield • Ridgecrest •
• Lompoc Tehachapi • • California City
Mojave • • Baker
• Palmdale • Barstow
Santa Paula • Twentynine Palms
Ventura • • Burbank
• Pasadena
Los Angeles • • San Bernardino
• Riverside
• Palm Springs
Vista • Indio •
San Diego • El Centro •

Tallahassee • Jacksonville
Gainesville • Palatka •
Ocala • Leesburg •
• Orlando
Tampa • • Lakeland
• Plant City
Ft. Pierce •
• Arcadia
Clewiston •
Pompano •
Miami •
Homestead •

flavor. Citrus grown in subtropical climates have higher sugar and acid ratio and are generally considered richer in flavor.

Some varieties that are naturally high in acid content become too acid when grown in California. In a semitropical climate, they will be sweeter and less acid. 'Temple' tangor and 'Seminole' tangelo are examples of varieties that are more flavorful when grown in Florida than when grown in California.

Humidity affects many fruit qualities including size, shape and juice content. Citrus grown in Florida are generally larger and more spherical than the same variety grown in a subtropical climate. The rind is thinner and somewhat more adherent.

Other fruit characteristics vary by climate. For example, the navel in the end of a navel orange is usually less pronounced in Florida than it is is in subtropical climates. The rind of the 'Temple' tangor is much smoother in Florida than it is in California.

Humidity affects juice content of citrus fruit. Florida citrus is generally juicier and more tender than the same variety grown in drier California.

Oranges are the main citrus crop in Florida. 'Valencia', 'Hamlin' and 'Parson Brown' are the dominant varieties. Large-fruited varieties such as the 'Washington' navel become too large in Florida. Varieties such as 'Hamlin', which are normally small in California, reach ideal market size when grown in Florida.

Lemons grow large and puffy in Florida and do not develop the acid content characteristic of California lemons. In addition, the cool climates of the California coast allow virtually continuous harvest of the lemon. These conditions make lemon growing most profitable in California.

Limes, on the other hand, are particularly well adapted to Florida's climate. The 'Persian' lime is the main crop. The 'Mexican' is also more successful in Florida than California.

CITRUS SUBCLIMATES OF CALIFORNIA

California's citrus climate is generally classified as subtropical. It is characterized by wet winters and dry summers. Similar climates are common in the Mediterranean region of Europe, another good citrus-growing area.

California's climate is influenced primarily by the Pacific Ocean. Furthermore, the medium-elevation coastal mountain range and the high-level Sierra Nevada range serve to isolate large areas from the heat and cold of the continental United States.

The Sierra Nevada forms a barrier from north to south that creates a climatic island. Areas of generally similar climate that normally run from east to west throughout the rest of the United States, run north to south in California. Elevation and proximity to the ocean are more significant in California than latitude.

California is divided into three citrus growing zones: the cool, moist, coastal area; warmer inland regions; the hot, dry desert. Each is different and illustrates the dramatic effect of heat and humidity on citrus.

Since the mid-1960's, the University of California has been recording citrus fruit differences at locations representing the three commercial zones. Two sites are in the coastal region, one in the warmer Central Valley, and one in the low-elevation, hot, dry desert.

Fruit characteristics were recorded for 'Washington' navel and 'Valencia' oranges, 'Eureka' and 'Lisbon' lemons, 'Dancy' and Satsuma mandarins, and 'Marsh' and 'Redblush' grapefruit. The study confirmed the significant influence of climate on citrus fruit characteristics.

Total heat is most important and primarily determines the date when fruit ripen. Citrus grown in the desert areas are usually first to ripen, then come those in the inland areas, and finally those in the cool coastal zone. Fruit grown closer to the coast hangs on the tree longer after maturity without deteriorating.

Heat can be the greatest limiting factor to the production of edible fruit. Simply stated, heat translates into sweetness, and sweetness means ripeness.

Some varieties have high heat requirements that can only be met in the warmest locations. Such is the case with grapefruit, at its highest quality in the California low desert, the Arizona desert, and Texas. Closer to the coast, grapefruit ripens much later and is commercially significant as an "off-season" fruit, but the quality is never as good as desert grapefruit.

Holding time on the tree is important with grapefruit. Near the coast, grapefruit can be left on the tree for months without deteriorating. They will gradually increase in sweetness. The homeowner can leave the fruit on the tree until its flavor is acceptable.

Acid citrus, such as lemons and limes, don't need to sweeten-up. They have low heat requirements and can be grown in cool locations. Cool coastal conditions accentuate the "everblooming" characteristics of lemons, resulting in several harvests a year. Under desert conditions, lemons ripen mainly in late fall and winter.

Heat and light are responsible also for the internal and external coloration of many distinctive varieties. 'Redblush' grapefruit only develops good red flesh with high heat. In cool areas, this red color fails to develop and it looks like the white-fleshed 'Marsh'.

Light has many effects on citrus. In their native habitat, many citrus types grow in the shade of taller plants. Grown in the shade, citrus leaves are fewer but larger and softer textured. Near Indio, in the California low desert, many citrus trees grow in the shade of date palms. This is only possible in areas with high heat and intense light. In cooler areas, the loss of heat and light in such a location would retard or prevent satisfactory production of most varieties.

Citrus fruit and bark are sensitive to sunburn. Under the intense light and heat of the desert, fruit quality may be adversely affected on the exposed south side of the tree due to sunburn. Exposed limbs should be protected.

University of California research shows that the fruit of most citrus varieties grow larger in the desert. Exceptions were the Satsuma mandarin and the 'Washington' navel, which are not usually recommended for desert conditions. Such varieties are adversely affected by high heat at blossom time. Fruit grown in desert climates will generally have less juice and a rougher rind than the same varieties grown under milder climates.

From our trip through tropical, semitropical, and subtropical citrus areas we can see the important role climate plays in citrus growth and production. An understanding of climatic effects helps you determine what cultural steps you must take for best results in your climate.

THE COOL WINTER ADVANTAGES

After reading the previous section on the effect of climate on citrus, you realize that there is an important association between the quality of citrus fruit and the latitude at which it is grown. Areas farther north or south of the equator produce better quality fruit. On the other hand, the farther you go away from the equator, the greater the threat of frost. Contending with the threat of frost is the price you pay for better tasting fruit. Commercial citrus growers know this well. In California, severe frosts have occured in 1913, 1922, 1936, 1962-63, and 1973. Although these frosts may have determined some boundaries, commercial citrus areas in California are increasing rather than decreasing.

Every ten years or so you will probably experience some frost damage on your citrus. But citrus trees are surprisingly resilient. A tree that loses all its leaves can come back strong and produce fruit the following year. Even if the tree is killed, the price of a new one is

Cold Tolerance

Mexican Lime   Lemon   Grapefruit   Pummelo

Most Tender

small compared to ten years of fruit production.

COLD HARDINESS

Different types of citrus show varying degrees of hardiness. Precise temperatures at which cold damages citrus cannot accurately be predicted. The citrus species listed from the most tender to the most hardy is: citron, lime, lemon, grapefruit and pummelo, sweet orange, sour orange, mandarin and kumquat. The tangelo and tangor hybrids rank about equal to the sweet orange in cold hardiness. The 'Meyer' lemon, Rangpur lime, calamondin, Satsuma mandarin and kumquat hybrids rank intermediate between the mandarin and kumquat species.

There are many variables that determine cold hardiness.
• Duration of cold. A few minutes below freezing is less damaging than an hour.
• Position of the fruit. Fruit well-covered by foliage is more protected from cold.

Tangelo   Orange   Mandarin   Meyer Lemon   Kumquat

Most Hardy

• The position of the tree in the garden is important. Is it near a warm wall or exposed to the open sky?

• Is there good air drainage? Cold air will fill low areas, damaging citrus planted there.

Some guidelines can be helpful. Citrus types are usually listed in order of increasing foliage hardiness. 'Mexican' limes and citrons are generally damaged at temperatures slightly below 32°F (0°C). Kumquats are usually considered hardy to 18° to 20°F (-8° to -7°C). Citrus fruit is usually less hardy than the foliage. Most fruit is damaged between 26° to 28°F (-3° to -2°C). Grapefruit is slightly more cold resistant. Mandarin fruit is more frost sensitive.

Young succulent growth and blossoms are the most tender, making late spring frosts most damaging.

THE MANY CLIMATES OF YOUR GARDEN

Every garden has many climates. There is the cool shade of a tree on a hot summer day, or the light and heat bouncing off a wall. Such pockets of unique climate are called *microclimates*.

In many areas, using microclimates is very important to the successful growing of citrus. Such an understanding, for example, may permit the fruit lover living in a cold "kumquat climate" to successfully grow more tender citrus such as lemons or oranges. Or a lemon grower, in a cool, coastal area, may try fruit with a higher heat requirement such as grapefruit.

A microclimate is a very small area, perhaps just a few square feet, where the climate differs from the general climate of the area. On a large scale, an oasis is

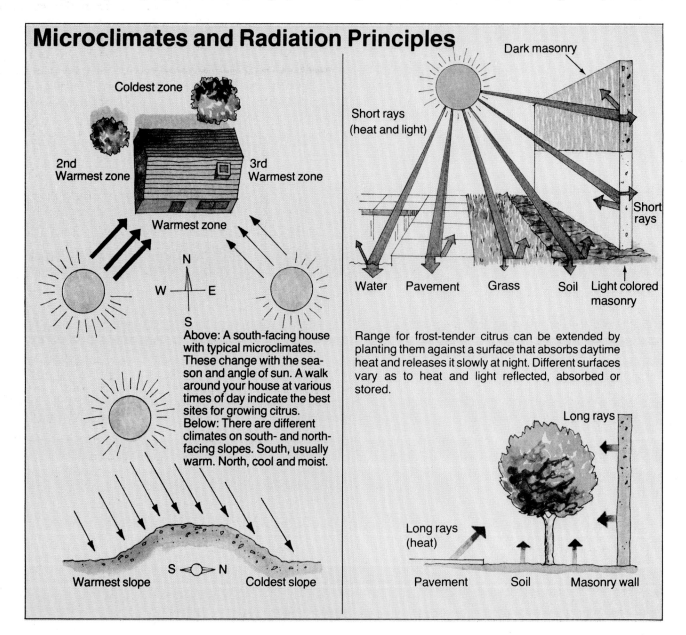

Microclimates and Radiation Principles

Coldest zone

2nd Warmest zone

3rd Warmest zone

Warmest zone

N
W — E
S

Above: A south-facing house with typical microclimates. These change with the season and angle of sun. A walk around your house at various times of day indicate the best sites for growing citrus.
Below: There are different climates on south- and north-facing slopes. South, usually warm. North, cool and moist.

Warmest slope   S ◆ N   Coldest slope

Dark masonry

Short rays (heat and light)

Short rays

Water   Pavement   Grass   Soil   Light colored masonry

Range for frost-tender citrus can be extended by planting them against a surface that absorbs daytime heat and releases it slowly at night. Different surfaces vary as to heat and light reflected, absorbed or stored.

Long rays

Long rays (heat)

Pavement   Soil   Masonry wall

a microclimate in a desert. A microclimate is the unique interaction of sun, heat, and cold with the physical objects of a small area. It involves processes such as absorption, radiation, reflection and air flow.

Most of what you need to know about microclimates is just a matter of common sense. To get a feel for them, take a walk around your house at various times of the day. The east side of the house is usually first to warm up in the morning. The north side will stay shady almost all day. The west side is warm and sunny in the afternoon. The south side has the most consistent sun all day. Obviously, these conditions are varied by movement of the sun, but they are just a few of the various microclimates around your home. Each microclimate affects plant growth in unique ways.

The time of year that you take this stroll around your home is important. Most people realize that the sun changes position in the sky from summer to winter. The summer sun is at a high angle and is therefore warmer. The winter sun is at a lower angle in the southern sky and is cooler. Since the sun is the major climate-controlling influence, the microclimates around your home will change with the seasons and the angle of the sun.

The south side of your house should be the warmest, especially in winter, provided, of course, that it is not shaded by tall trees. Citrus grown there will receive the most sun and warmth.

People living in hilly areas ought to be familiar with the different climates on south and north-facing slopes. A south slope is usually warm and dry while the north slope is cool and moist. Each supports different types of plant life.

You can extend the range of some of the more frost-tender citrus by an intelligent use of microclimates. For example if you live in a cold kumquat climate you might be able to grow oranges or even lemons by planting them against a surface that absorbs heat during the day and releases it slowly at night. The night temperatures up against a warm south-facing wall may be several degrees higher than in an open area just a few feet away.

ABSORPTION, RADIATION, REFLECTION AND YOUR MICROCLIMATE

Different surfaces vary as to the amount of heat and light reflected, absorbed or stored. Light-colored surfaces reflect more heat and light; dark surfaces absorb them. Heat stored will be released as air temperatures around the surface begin to cool and the light hitting these surfaces decreases.

Such processes affect ground surfaces as well as walls. Water is the best heat storage reservoir. It absorbs as much as 95 percent of the heat that reaches it through sunlight and then radiates it slowly back to the surrounding area at night. Various types of stone, cement or masonry absorb about 50 to 60 percent of the heat depending on their color and density. Average soil absorbs about 30 percent with cultivated soil dropping to about 20 percent because of air spaces. Grass and leafy mulches absorb relatively little heat because of air spaces and rapid cooling.

To put the importance of radiation, reflection and absorption into proper perspective, let's consider an example. Let's say you live in a moderate climate usually reserved for lemons, but you want to grow citrus with high heat requirements, such as grapefruit. By planting between a cement slab and a hot south wall, you may be able to reflect enough heat to grow excellent grapefruit. This is how microclimates are used to expand your citrus climate.

Planting in raised beds, in which the soil is quick to warm up, and espaliering against a south wall are other good ways to maximize heat.

WIND

Wind can increase heat loss of citrus trees and also damage fruit. By the use of various screens or buffers, you can redirect desiccating or damaging winds while at the same time reflecting more heat onto the tree.

COLD AIR DRAINAGE

Cold air seeks the lowest level just as water does. As the ground cools after the sun goes down, the cooler air follows the natural slope of the watershed, flowing down valleys and settling in the lowest spots. That is why flat valley floors are usually colder than sloping land along the valley sides. This movement of cold air helps account for the thermal belts that exist on hillsides in many parts of the country. Rather than damming up, the cold air drains away, leaving some areas frost free.

If you understand how cold air flows in your garden, you can often save tender plants from frost damage. First of all, avoid planting in low spots where cold air accumulates. Provide air drainage whenever possible. Because a fenced yard can trap cold air, simply opening a gate will often allow for drainage.

FROST PROTECTION

The time will probably come when temperatures drop so low that you'll need to provide your citrus with some frost protection.

The first step is predicting when a frost will occur or when temperatures will drop to a point damaging to your citrus. Most areas have weather radio stations that forecast frosts. Strategically placed thermometers are also helpful. Some long-time gardeners say they can "taste" a frost coming.

Frosts occur when there is maximum heat radiation from the ground into the atmosphere. This usually happens on cold, clear, still nights. Clouds block infrared radiation and prevent frosts.

The University of California recommends the following for frost protection: "Wrap a thick insulating material, such as fiberglass building material, or two or three layers of corn stalks or newspapers upright around the tree trunk, and tie them firmly in place. Trunks of young trees are most susceptible. Leave some foliage exposed so the tree can function normally. To prevent brown rot gummosis disease, spray or paint the trunk and main lateral shoots with a neutralized copper (Bordeaux) mixture before wrapping.

"Build a stout frame around the tree. Cover, using burlap or clear plastic, with removable top and removable south side. Neither the burlap nor clear plastic should remain over the plant during sunny weather. Do not merely drape a cover over the tree. Foliage touching the cover will be damaged by the cold.

"For prized plants, a heat lamp or a couple of 150 watt light bulbs located in the central area of the shelter or in the tree crotch may add just enough heat to prevent damage. Where excess water may be success-

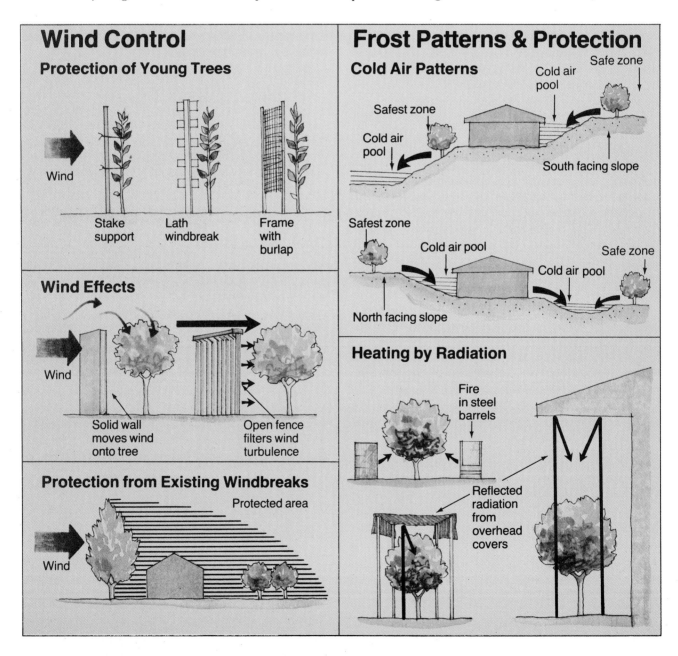

Wind Control
Protection of Young Trees

Wind

Stake support | Lath windbreak | Frame with burlap

Wind Effects

Wind

Solid wall moves wind onto tree

Open fence filters wind turbulence

Protection from Existing Windbreaks

Protected area

Wind

Frost Patterns & Protection
Cold Air Patterns

Safe zone
Cold air pool
Safest zone
Cold air pool
South facing slope

Safest zone
Cold air pool
Cold air pool
Safe zone
North facing slope

Heating by Radiation

Fire in steel barrels

Reflected radiation from overhead covers

fully drained away, low head sprinklers placed on opposite sides of a tree applying about a half gallon a minute per tree throughout the night may provide an added 2° to 3° (-11° C). The sprinklers should be turned on before air temperature lowers to 32° F (0° C). and should remain on until the temperature rises.

CAUTION—Do not sprinkle water over trees because a buildup of ice can cause severe limb breakage. If the air is unusually dry, evaporation of sprinkler water may cool rather than provide heat.

"After the third year it is usually no longer necessary to insulate tree trunks. Other methods suggested for young trees can also be used on older trees."

INCREASING HARDINESS

Besides the use of microclimates and physical protection, there are a number of cultural practices that can increase effective hardiness.

As stated in the chapter entitled "Growing Citrus," citrus growth is cyclic. It goes through regular periods of reduced growth in which foliage hardens off and there is little or no new growth. Citrus reaches maximum hardiness during this period, so it makes sense to avoid stimulating growth prior to the cold season.

Under a proper fertilizing program in which nitrogen is withheld beginning in midsummer, citrus growth should gradually slow by fall. Excess nitrogen in late summer and fall can cause growth flushes of very frost-tender foliage. Such trees will be less hardy.

ROOTSTOCKS

The rootstock on which a variety is grafted also has a profound influence on its hardiness. Trifoliate orange used as a rootstock increases hardiness. The same is true of the citrange rootstock although to a lesser degree.

Both of these citrus relatives are deciduous although the citrange is leafless for a much shorter period than the trifoliate orange. They prolong dormancy of the scion variety and should therefore be the preferred rootstocks in colder climates.

CHOOSE THE RIGHT VARIETY

Give careful thought to variety selection if you want to grow citrus in colder areas. Obviously, the hardiest types are the best. Kumquats, kumquat hybrids and the Satsuma mandarins are the tough ones. Many of the citrus relatives are also very hardy.

Early ripening varieties, like the Satsuma, are also good choices, because the fruit can often be picked before damaging frosts.

Check the variety descriptions in the chapter, "Which Varieties Will You Choose?"

Citrus Climates of the West

ZONE 1—SOUTHERN CALIFORNIA COASTAL

Within this zone lie some of the most equitable climates in the United States. It comprises an area from San Diego north to San Luis Obispo, and is governed primarily by ocean influences. The cities of Santa Barbara, Ventura and areas of Los Angeles lie within this zone.

Summers are usually cool and pleasant with temperatures moderated by coastal fog. However, the inland extremities of this zone begin to resemble Zone 2.

Cold winter temperatures are not usually a threat to citrus in this zone. Occasionally (once in ten to twenty years), frigid arctic air may sweep down the valleys with devastating effect. Also, some low, wind protected areas within this zone can become quite cold. Citrus grown here do not become as dormant as in typically cooler zones, and can be damaged by a sudden spell of temperatures not normally considered threatening.

Parts of Zone 1 may be relatively frost free. Many gardeners in these areas enjoy the pleasures of tender tropical and subtropical plants. It is often called the Avocado or Banana Belt.

Wind is an adverse factor when citrus trees are exposed directly to the ocean. Wind protection in such locations is a must. In addition, parts of Zone 1 are subject to hot, drying winds known as *Santa Anas*. They usually occur in winter or early spring and can greatly damage citrus trees and fruit.

Zone 1 is important commercial citrus country. Lemons are the main crop. Ventura County is the largest lemon producing area in the state. The cool, moderate climate stimulates fruiting of the 'Eureka' lemon almost the year-around. Fruit is harvested three to five times a year.

The fruit grower in Zone 1 has an opportunity to grow the most frost-tender citrus such as citrons and limes. But he must realize that the higher temperatures required to ripen many citrus types may limit what he can grow. Varieties with low heat requirements are the most satisfactory. Acid fruits such as lemons and limes are ideal. Several miles inland in wind protected areas, 'Valencia' oranges and occasionally navel oranges may do well. With optimum use of microclimates and by allowing fruit to remain on the tree for as long as possible, success can be achieved with varieties usually reserved for warmer climates.

Compared to other zones, fruit in Zone 1 ripens last.

ZONE 2—INTERMEDIATE REGIONS OF SOUTHERN CALIFORNIA

This zone includes the warmer areas inland from the coast. It encompasses regions east of Los Angeles typified by the cities of Riverside, Upland, Corona and San Bernardino, as well as areas closer to the coast such as the San Fernando Valley and Pasadena. The Ojai Valley, inland from Ventura, is also in Zone 2.

The climate in this zone is controlled to varying degrees by two factors: cool coastal air and hot desert air. The summers are generally warm and dry with cool nights and occasional coastal fog. Winters are pleasant because of the influx of warm air from the nearby desert. Occasional hot Santa Ana winds can damage citrus fruit.

Winter low temperatures depend on elevation and slope of the land. In certain upland areas where cold air drains away, leaving a warmer thermal belt, less hardy citrus can be grown. In lower areas, where cold air settles, winter temperatures dip into the twenties more often and hardier varieties should be planted.

At one time this was the most important commercial citrus area in California. The University of California at Riverside is a major center of research on citrus. Today, Zone 2 continues to feel the pressures of urbanization, but sizable acreages of oranges and lemons still remain. Zone 2 is also important for production of summer grapefruit.

Homeowners in this area can raise almost any variety of citrus, as long as its specific cultural requirements are met. There is plenty of summer heat to sweeten fruit, while protection from winter cold can easily be provided if the site is chosen carefully.

ZONE 3—CALIFORNIA AND ARIZONA DESERT REGIONS

The cities of Palm Springs, Indio, Tucson and Phoenix are included in this zone.

Summers are long and harsh in the desert. Wind and extreme heat often tax citrus adaptability. Still, this area is excellent for many types of citrus, especially those with high heat requirements.

Citrus grown in the desert ripen early which makes this zone an important growing area commercially. Grapefruit find desert conditions superb and yield good crops of sweet, high quality fruit. The same is true for 'Valencia' oranges, and many varieties of mandarins, lemons, tangelos and tangors.

Home citrus growers, however, must realize that conditions are different in the desert than in other citrus areas. Heat stress often causes sunburned fruit, particularly on the south side of the tree. When citrus bark is exposed because of excessive leaf drop or pruning, it must be protected against the intense desert sun.

Water and soil conditions are often poor. Salty, alkaline soil may limit citrus growth.

Winter temperature extremes sometimes cause citrus losses. Temperatures occasionally fall below freezing several nights during a cold year, and frost protection is needed. At higher elevations, such as around Tucson, winters are frequently colder. There are climatic differences between the California desert and the Arizona desert as well as between Phoenix and Tucson. For this reason, there are additional descriptions of the Arizona climates with the climate map.

ZONE 4—CALIFORNIA CENTRAL VALLEYS

This large zone includes the San Joaquin and Sacramento Valleys of central California. Although the northern tip of this zone is more than 400 miles from the southern end, areas of similar elevation have remarkably similar climates. The reasons for this are explained in "Citrus Subclimates of California," page 29. However, since the northern and southern parts of this zone have developed differently and because of some climate variations in the southern Sacramento Valley, we will discuss them separately.

The San Joaquin Valley includes the cities of Bakersfield, Visalia and Fresno. Summers are hot and sunny. Winter temperatures depend largely on the slope of the land and the downward flow of cold air. A thermal belt exists along the edge of the valley which is suitable for many types of citrus. In fact, at elevations of 500 to 700 feet, damaging frosts are relatively infrequent.

The protective tule fog favors citrus growing in this area. In winter, conditions create this dense, low-lying fog which often acts like a thermal blanket and prevents frosts. If the tule fog lies below the thermal belt, higher elevations may actually be colder than the valley floor.

This cool, moist, fog is a major difference between Zone 4 and Zones 2 and 3 where winters are warmer and a sudden cold spell can be very damaging. Citrus grows very slowly in tule fog areas which also makes the fruit less susceptible to frost damage. However, because of the lower average temperatures on the valley floor, hardier or early ripening varieties are usually planted.

Commercially, the San Joaquin Valley is the most important citrus producing area in California. There are many acres of navel oranges as well as 'Valencia' oranges, mandarins, mandarin hybrids and lemons.

For the home citrus grower, this area offers an opportunity to grow many kinds of citrus. Avoid the tender limes and citron, but the 'Tarocco' blood orange is better adapted here than anywhere else in the state.

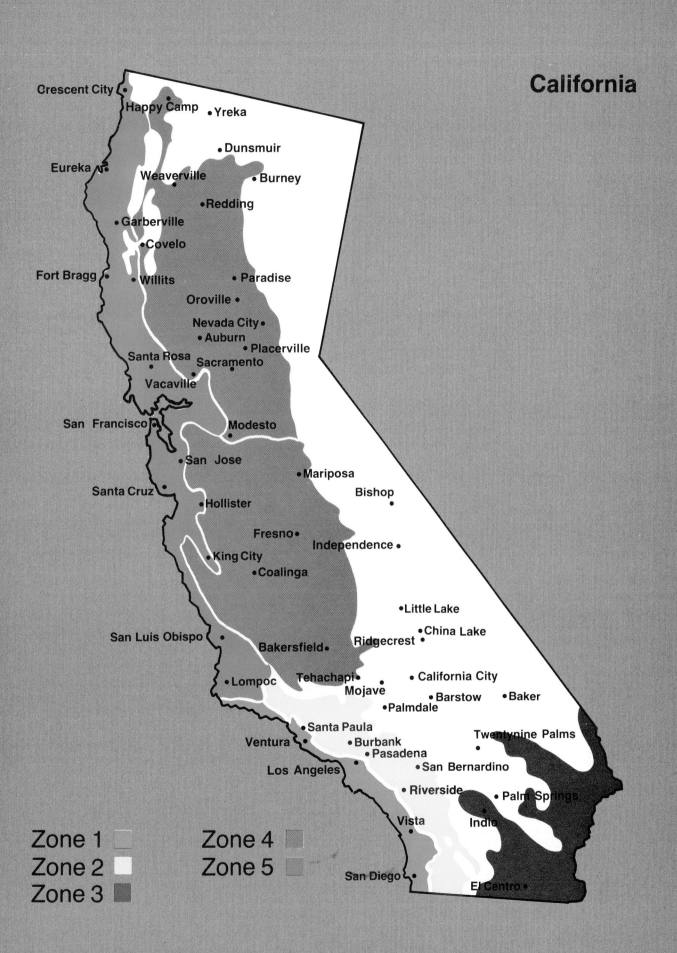

California

Crescent City

Happy Camp • Yreka

• Dunsmuir

Eureka

Weaverville • Burney

• Redding

Garberville

• Covelo

Fort Bragg

• Willits • Paradise

Oroville •

Nevada City •
• Auburn

Santa Rosa • Placerville

Sacramento

Vacaville

San Francisco

• Modesto

San Jose • Mariposa

Santa Cruz Bishop

Hollister

Fresno • Independence •

King City

• Coalinga

San Luis Obispo • Little Lake

• China Lake

Bakersfield • Ridgecrest

Lompoc Tehachapi • • California City

Mojave • Barstow • Baker

Palmdale

• Santa Paula Twentynine Palms

Ventura • Burbank
• Pasadena

Los Angeles • San Bernardino

• Riverside
• Palm Springs

Vista

Indio

San Diego

El Centro •

Zone 1   Zone 4

Zone 2   Zone 5

Zone 3

Citrus climates **37**

The Sacramento Valley and surrounding areas include the cities of Sacramento, Marysville, Oroville, Paradise and Chico. Like the San Joaquin Valley, they have warm summers and cold winters often dominated by tule fog. Varieties that can be grown depend largely on the slope and exposure of the planting site. Rising from the valley floor on higher, sloping land there exists a thermal belt used in the past for commercial citrus production.

In areas around Oroville and Auburn are small commercial orchards of navels and Satsuma mandarins. "U-pick-it" Satsuma mandarin orchards are located on the "49er Fruit Trail" near Auburn. Many more acres that once produced fruit for market have been abandoned or subdivided. Some homeowners have preserved remnants of these orchards to create ornamental landscapes around their homes.

The northern portion of Zone 4 differs from the southern section because occasional ocean air cools off normally hot summer days in the southern Sacramento Valley.

Early and hardy citrus varieties are the most successful in the Sacramento Valley area. 'Washington' navels and Satsuma mandarins are most important. With good air drainage and protection, many other less hardy varieties are worth trying.

ZONE 5—NORTHERN CALIFORNIA COASTAL VALLEYS

This large zone contains many different climates all influenced to varying degrees by the Pacific Ocean. It includes the San Francisco Bay area, San Jose, Monterey, and south, as well as north to Santa Rosa, Cloverdale and Napa.

In summer, the weather may be cool, windy and foggy near the coast in San Francisco, but rather warm in inland areas such as St. Helena or Gilroy. Although summer fog buffers most of these inland areas from the high temperatures of the Sacramento Valley, summer heat waves take temperatures into the high nineties and low one-hundreds.

Winters are quite varied. Near the coast in San Francisco and in thermal belts in many valleys, frosts are infrequent. On the other hand, the low lying inland areas often suffer temperatures below freezing. Site selection, air drainage and winter protection thus become very important considerations.

Although relatively unimportant compared to southern California, commercial citrus production has a long history in this area. In northern areas around Cloverdale, 'Washington' navels still grow in small parcels. Satsuma mandarins have also proved their worth. The annual Cloverdale Citrus Fair celebrates the fruits which have played an important role in its history.

In northern cities like St. Helena, Santa Rosa and Napa, old plantings of grapefruit, oranges and lemons can still be seen on the streets and around some older homes.

Many citrus nurseries are located in this zone. Because the area is relatively virus free, two of the major producers of dwarf citrus have growing grounds here.

Many kinds of citrus are grown in this area. In the San Francisco bay area, close to the coast, varieties with low heat requirements are preferred. 'Trovita' orange, 'Robertson' navel, lemons, 'Bearss' lime, 'Sanguinelli' blood orange and Satsuma mandarins are popular. Strategic planting in warmer microclimates can also bring success with pummelos, other mandarins and even grapefruit.

Farther inland there is usually plenty of summer heat, but winter cold can be a problem. Early navels and Satsuma mandarins do well. Hardy citrus such as kumquats, 'Meyer' lemons and grapefruit are desirable. With protection and good site selection there are many possibilities.

SPECIAL NOTES ON ARIZONA

The information on Zone 3 on page 36 applies to the mid-altitude Arizona desert. Additional information applying only to this Arizona area, designated Zone 3 and 3A, is found below. Zone 3B is for the low-altitude Arizona desert.

MID-ALTITUDE ARIZONA DESERT—ZONE 3-3A (Elevation 2,000-4,000 feet)

Tucson's growing season is about 242 days long with an average last date of spring frost on March 20. Safford and Wickenburg have a growing season about three weeks shorter. Sour orange, xylosma, pineapple guava, Texas ranger, olive and star jasmine are representative plants.

The mild winters here do not meet the cold requirement of many deciduous fruits, flowering trees and shrubs. On the other hand, the tender citrus must be protected from the hard frosts which occur here. September, October and November are ideal months for planting. This allows time for plants to become well-established in the new location before the onset of summer heat.

LOW-ALTITUDE DESERT—ZONE 3B (Elevation 100-2,000 feet)

Growing seasons here are long, ranging from 302 days in Phoenix to 340 days in the Yuma area. The average minimum winter temperature recorded for this zone is around 36-37° F(2-3° C). However, temperatures occasionally dip below 20° F(-2° C). In summer, averages maximum temperatures are near 102° F(39° C). Annual rainfall is ten inches or so. The fall months of September and October signal the beginning of the planting year.

Arizona

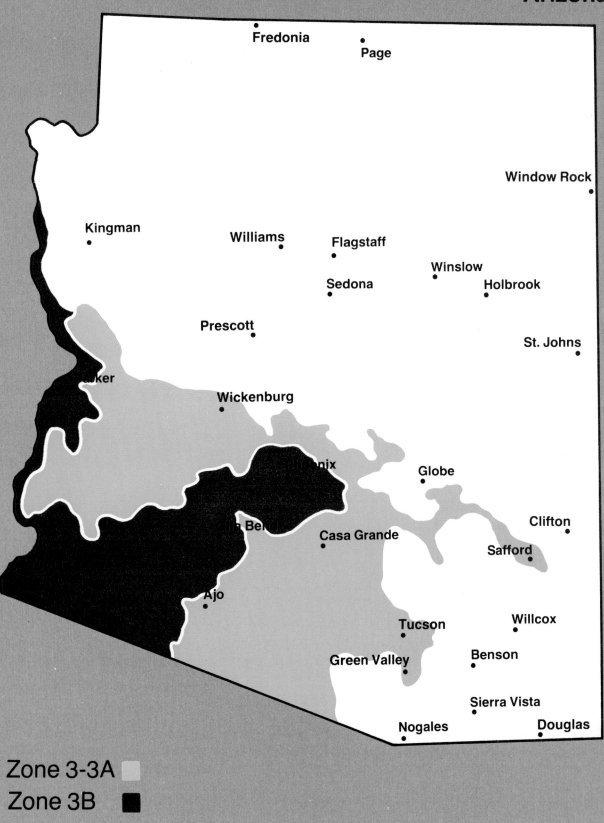

- Fredonia
- Page
- Window Rock
- Kingman
- Williams
- Flagstaff
- Winslow
- Sedona
- Holbrook
- Prescott
- St. Johns
- arker
- Wickenburg
- nix
- Globe
- Clifton
- a Ben
- Casa Grande
- Safford
- Ajo
- Willcox
- Tucson
- Benson
- Green Valley
- Sierra Vista
- Nogales
- Douglas

Zone 3-3A
Zone 3B

Pummelo   **Grapefruit**   **Orange**

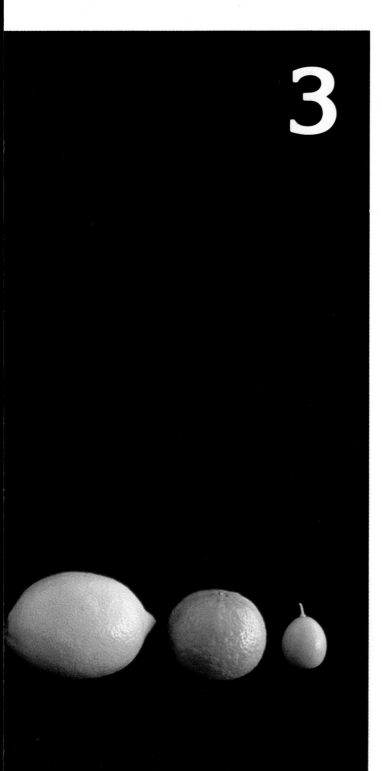

3

Which Varieties Will You Choose?

Lemon   Mandarin   Kumquat

Citrus vary in size, color, shape and flavor. Some are the same today as centuries ago, while others are new. Many are worth your consideration.

There are many factors to consider when you select a citrus variety. Probably the most important is its climate adaptation. Reading the chapter entitled "The Effects of Citrus Climate" will be helpful. The information of this chapter covers 40 varieties of citrus. Whether the information is important to *you* depends on your location, personal tastes and how the tree will be used.

For instance, if you live in a warm inland area rela-

The citrus family is usually represented by various shapes, sizes and colors. Flavor, juice, peel and hardiness are other important factors in variety selections.

tively free of frosts, it makes little difference that the 'Skaggs Bonanza' or 'Robertson' navel matures two weeks before the 'Washington' navel. But if you live where early frosts are a threat, the fact that these are early ripening varieties may make a big difference. For the person growing citrus for fresh juice, ease of peeling and seed content means very little. But someone who likes to peel the fruit and eat it, such things are important and will tip the scales in favor of one variety over another.

There is a diversity in the citrus family that is rarely matched among plant groups. Harvest periods can be blended to provide almost 12 months of fresh fruit. With indoor-outdoor container culture, citrus is adaptable to most climates. And, their wide range of ornamental qualities fills many landscape needs.

Think carefully about placement of the tree in the garden. The blood orange may need a special microclimate against a warm, south-facing wall. Or you may desire the fragrance and bright foliage of an everbearing 'Meyer' lemon on the patio.

Realizing that few people will be able to grow all the varieties of citrus they want, we have included a supermarket buyer's guide for the commercially important fruit types. You will find these at the end of each section. Oranges, mandarins, tangelos, tangors, lemons, limes and grapefruit are discussed in detail with variety availability and tips on buying the best fruit.

The eye appeal of citrus is something we have concentrated on throughout this book. On the following pages we let the fruit and trees speak for themselves. There are many photographs, at least one of each variety. The ultimate experience, however, is yet to come; when you taste the fruit off your own tree.

These 40 citrus varieties were collected in January from one location, the University of California Lindcove Field Station. Lindcove is in the California Central Valley, Zone 4. Shape, color and size will vary slightly in other climates. Time of optimum maturity also varies. For instance, here 'Redblush' and 'Marsh' grapefruits differ only slightly. Grown in the California or Arizona desert, Zone 3, 'Redblush' will develop more skin and flesh coloration.

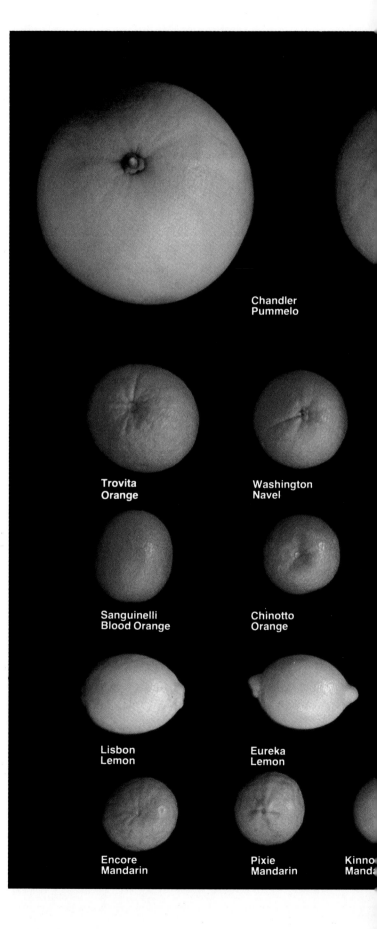

Chandler Pummelo

Trovita Orange

Washington Navel

Sanguinelli Blood Orange

Chinotto Orange

Lisbon Lemon

Eureka Lemon

Encore Mandarin

Pixie Mandarin

Kinno Manda

Variety Size and Color Comparison

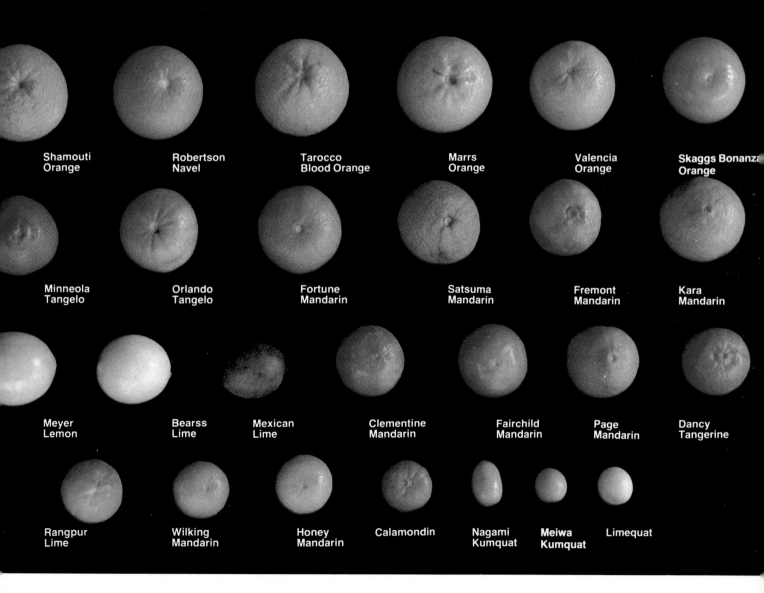

Reinking
Pummelo

Marsh
Grapefruit

Red Blush
Grapefruit

Etrog
Citron

Shamouti
Orange

Robertson
Navel

Tarocco
Blood Orange

Marrs
Orange

Valencia
Orange

Skaggs Bonanza
Orange

Minneola
Tangelo

Orlando
Tangelo

Fortune
Mandarin

Satsuma
Mandarin

Fremont
Mandarin

Kara
Mandarin

Meyer
Lemon

Bearss
Lime

Mexican
Lime

Clementine
Mandarin

Fairchild
Mandarin

Page
Mandarin

Dancy
Tangerine

Rangpur
Lime

Wilking
Mandarin

Honey
Mandarin

Calamondin

Nagami
Kumquat

Meiwa
Kumquat

Limequat

How to Use This Chapter

Each citrus group begins with an introduction explaining unique adaptations, historical significance and cultural specifics.

Variety descriptions follow, broken down by characteristic.

Tree Size and Characteristics: A brief discussion of tree characteristics that will influence its ornamental quality and location in the garden. This includes relative, rate of growth, habit of growth, position of the fruit in the canopy, (inside, intermediate, or outside) color and density of foliage and relative thorniness.

Fruit Size: In relation to other fruits of its type.

Fruit Color: Peel color and, if distinctive, pulp color.

Ease of Peeling: With the obvious exceptions such as lemons, limes and citrons, almost all citrus can be peeled without too much difficulty. But some are easier to peel than others. Puffiness of the rind can be a sign of overripeness rather than an easy-to-peel fruit.

Seeds: Not a good criteria on which to base varietal selection. Some of the very best tasting varieties have many seeds. Seediness varies even within fruit from the same tree. See the section of this book on "Growing Citrus" for information on why some varieties have many seeds and others none.

Flavor and Juiciness: Flavor is one of the hardest fruit qualities to measure. It will vary a great deal due to climate and yearly weather fluctuations as well as the maturity of the tree when it is picked. Fruit technicians use indicators such as sugar-to-acid ratio to measure flavor, but the list of descriptive terms used is long and technical. For this reason, we have relied on the University of California descriptions for both flavor and juiciness. We have relied on personal experience where it seemed helpful.

Holds on the Tree: This refers to length of time the fruit can be stored on the tree after ripening, without substantial loss of quality. This varies a good deal according to location (generally longer in cool climates) and weather, but it is still the best way to store fruit. As a general rule, varieties listed as "excellent" can hang on the tree for at least three months after ripening. Some, such as 'Valencia' oranges and grapefruits, can last as long as five to six months. "Good" is two to three months, "Fair" from one to two months, and "Poor" less than one month.

Harvest Period: Based on the climate zones, these timetables are elastic and should only serve as guidelines. Year to year weather differences and microclimates within zones and around the home play an important role as to when fruit can be picked.

If a variety is not recommended for a particular zone, no harvest period is given. Some varieties yield irregularly in certain zones mentioned. This is in the descriptions, but we have still given harvest periods. Warmer areas farther inland will have earliest fruit for each zone.

Harvest periods go hand-in-hand with the ability of the fruit to hang on the tree after ripening. A longer holding time means a longer harvest period.

The best way to tell if fruit is ripe is to pick one and taste it. If it is not sweet enough, wait a while longer; if it is dry, it has been on the tree too long. Use pruning shears to harvest the fruit rather than pulling them off. Clip the stem with a smooth, close cut. Try not to cut or damage the fruit. Cuts and bruises speed deterioration after harvest.

General Comments – Varietal specifics on adaptation, ornamental qualities, origin and historical and commercial importance.

Sample Variety Chart

ENCORE

Tree Size: Medium.
Tree Characteristics: Upright, spreading habit. Moderate growth rate. Few or no thorns. Fruit position—intermediate.
Fruit Size: Medium.
Fruit Color: Orange with darker orange spots.
Ease of Peeling: Excellent.
Seeds: Many.
Holds on Tree: Good.
Flavor and Juiciness: Rich and juicy.

Zones		Harvest Period
S. Cal. Coastal	1	May - July
S. Cal. Inland	2	Mid Apr. - Mid July
Cal. & Ariz. Deserts	3	Mid Mar. - June
Cal. Central Valleys	4	Mid Apr. - Mid July
N. Cal. Coast. Valleys	5	May - Mid Aug.

'Encore' is a cross between a 'King' and 'Mediterranean' mandarin. It is a very valuable late-maturing variety which bears colorful speckled fruit of excellent flavor. Recently introduced by the University of California at Riverside, it is a heavy producer on an open tree. It does have a tendency to bear in alternate years.

Sweet Oranges

The most important member of the citrus family

Sweet oranges are by far the most important citrus crop in the world. Taste easily distinguishes sweet oranges from sour or bitter oranges. The bitter flavor of the sour orange is most obvious. They are discussed in detail on page 59.

Historically, oranges are believed to be native to northeastern India and parts of Burma and China, and were introduced into Europe as early as 200 to 300 A.D.

Oranges reached continental America in 1518 and Florida in 1565, and were introduced into Arizona by missionaries around 1707. It wasn't until after 1769, however, when Franciscan missionaries brought the sweet orange to California that they became an important food crop.

Since that time, oranges have played an important role in the development of the western United States. For a more detailed history, see page 5.

Today, estimates of commercial production indicate that there will be a total of almost 38 million tons of oranges grown world-wide in 1980. For California, 1978 state crop reports showed 183,640 bearing acres. Of all California fruits and nuts, only grapes and almonds are given more total acreage than sweet oranges.

Commercial orange production in California began with the introduction of the 'Washington' navel in 1873. The original tree can still be seen growing in Riverside. In the beginning, and for many years there-

Sweet oranges have a juicy fresh flavor and are a prime source of vitamin C. Thin skinned 'Valencia' is the most popular juice orange; navels are the most popular for eating. Choose sweet oranges that are firm and heavy.

From left: Four favorite oranges. 'Trovita', 'Robertson' navel, 'Valencia' and 'Washington' navel.

after, the Los Angeles basin was the most important citrus area in the state. It was only after the postwar urbanization in the late 1940's that production began to decline.

The climate of Los Angeles and the surrounding area proved ideal for growing 'Valencia' oranges and they became the main crop. The San Joaquin Valley became the major growing area for navel oranges.

In 1955, a total of approximately 98,000 acres throughout the state were planted in 'Valencia' oranges. More than 80,000 of these were in the southern California area around Los Angeles including Orange, Ventura, San Diego and Riverside counties. Navel oranges totaled 64,000 acres with more than 50 percent of that in the same southern California area.

By 1975, however, total 'Valencia' acreage had decreased to about 85,000 acres with only slightly more than half in southern California. Navels had grown to a total planting of about 125,000 acres with nearly 80 percent in the San Joaquin Valley.

Ventura, San Diego, Orange and Riverside counties still have large acreages of 'Valencia', but urbaniza-

Secondary fruit forms the navel in a navel orange.

From left: 'Hamlin', 'Marrs', 'Diller' and 'Pineapple'. All are sold in Arizona under the name Arizona Sweets.

tion and the price of land continue to reduce the number of groves.

Navel Oranges—You may wonder what causes the navel in a navel orange. Fruit specialists tell us the navel is the development of a secondary fruit at the end of the main fruit. This secondary fruit is the odd, top-shaped portion that is so evident when a navel orange is peeled and separated. As the second fruit enlarges, it causes the navel to enlarge.

As a group, navel oranges are known for their crisp, rich flavor, and ease of peeling and separation. They are among the finest table fruits, and certainly the standard of excellence among sweet oranges. They grow best and develop their best flavor under a narrower range of cool climatic conditions than other oranges, but the fruit ripens early enough to be picked before damaging cold weather occurs. This means that they can be grown successfully in colder areas.

In California, the most important orange is the navel, and the most important variety is the 'Washington', both commercially and for the home gardener. Bud sports (mutations or changes which have appeared on only one or two branches of a tree) of the 'Washington', such as 'Trovita', 'Robertson' and more recently 'Skaggs Bonanza' and 'Tule Gold', have become popular garden varieties.

Common Oranges—Characteristics of the common oranges are familiar to most people. Compared to navel and pigmented (or blood) oranges, common oranges comprise two-thirds to three-fourths of all oranges grown in the world. Besides the popular 'Valencia', widely grown in Florida and California, common oranges also include the commercially grown 'Parsons Brown', 'Hamlin' and 'Pineapple' of Florida, and the 'Jaffa' or 'Shamouti' of Israel.

Arizona Sweets—Several varieties of oranges are sold in Arizona under the name Arizona Sweets. They include 'Marrs', 'Pineapple', 'Diller', 'Hamlin' and 'Trovita'. These varieties are known by their variety names and are also popular in Texas and the Gulf Coast states. They are the most dependably bearing sweet oranges and avoid some of the alternate bearing tendencies of 'Valencia' in these areas. To find out which variety you are buying under this label, ask your nurseryman.

Washington Navel

Tree Size: Medium.
Tree Characteristics: Round-topped, slightly drooping habit with dense, dark green foliage. Fruit position is intermediate.
Fruit Size: Large.
Fruit Color: Deep orange.
Ease of Peeling: Excellent.
Seeds: Usually none.
Holds on Tree: Fair to good.
Flavor and Juiciness: Rich flavor. Moderately juicy.

Zones		Harvest Period
S. Cal. Coastal	1	Mid Jan. - Mid May
S. Cal. Inland	2	Dec. - Mid May
Cal. & Ariz. Deserts	3	Mid Nov. - Dec.
Cal. Central Valleys	4	Mid Nov. - April
N. Cal. Coast. Valleys	5	Dec. - May

The 'Washington' navel, one of the most important orange varieties, is the standard eating orange. Excellent in flavor, it is easy to peel and easy to separate into segments. Its introduction into California in 1873 from Brazil marks the beginning of the citrus industry in the western United States.

Commercially, it has a narrower range of adaptation than the 'Valencia'; it fruits poorly in the high heat of the desert and the low heat of coastal regions. However, because it ripens early, often before heavy frosts, it can be grown in colder areas than the 'Valencia'. This makes it a useful variety for cooler climates such as parts of northern California.

Because the 'Washington' navel can produce fruit where winters are colder, it plays an important role in commercial citrus production.

Trovita

Tree Size: Tall.
Tree Characteristics: Rather vigorous, upright. Very attractive, dense, dark green foliage. Fruit held toward inside of tree.
Fruit Size: Slightly smaller than 'Washington' navel.
Color: Orange.
Ease of Peeling: Good.
Seeds: Few.
Holds on Tree: Good.
Flavor and Juiciness: Pleasantly sweet and juicy.

Zones		Harvest Period
S. Cal. Coastal	1	Feb. - Mid June
S. Cal. Inland	2	Feb. - April
Cal. & Ariz. Deserts	3	Jan. - Feb.
Cal. Central Valleys	4	Feb. - April
N. Cal. Coast. Valleys	5	Feb. - Mid June

Although 'Trovita' is considered a seedling of the 'Washington' navel, it does not have the characteristic navel. Its major advantage over its parent is a wider range of adaptation. It fruits well under desert conditions and near the coast. 'Trovita' is especially prized in the San Francisco Bay area where it develops excellent flavor without high heat. At the same time, it fruits heavily in the Arizona and California deserts.

The fruit of 'Trovita' is slightly smaller than the 'Washington' navel, but juicier. It is considered a fine juice orange. It tends to bear heavily in alternate years.

Skaggs Bonanza

Tree Size: Medium.
Tree Characteristics: Similar to 'Washington' but with a smaller and more dense round head. Fruit position is intermediate.
Fruit Size: Medium-large to large.
Fruit Color: Orange.
Ease of Peeling: Excellent.
Seeds: Usually none.
Holds on Tree: Fair to good.
Flavor and Juiciness: Rich and sweet. Moderately juicy.

Zones		Harvest Period
S. Cal. Coastal	1	Mid Dec. - Mid May
S. Cal. Inland	2	Mid Nov. - Mid April
Cal. & Ariz. Deserts	3	Nov. - Mid Feb.
Cal. Central Valleys	4	Mid Nov. - Mid April
N. Cal. Coast. Valleys	5	Mid Dec. - Mid April

'Skaggs Bonanza' is a recently introduced, patented variety that is gaining commercial importance. It is just beginning to become available in retail nurseries on dwarf rootstocks. A bud sport of 'Washington' navel discovered in California, it produces more fruit which ripens about two weeks earlier than the 'Washington', bears fruit at an earlier age and has a smaller, more dense growth habit. It is a better choice than the 'Washington' navel in areas where frost is a concern because the fruit can be picked earlier.

'Tule Gold' and 'Atwood' are two other navel varieties that have characteristics similar to 'Skaggs Bonanza'. If these are available at your nursery, they are also good selections.

Robertson Navel

Tree Size: Small to medium.
Tree Characteristics: Slow growing. Fruit borne in clusters near outside of tree.
Fruit Size: Medium-large.
Fruit Color: Orange.
Ease of Peeling: Excellent.
Seeds: None.
Holds on Tree: Good.
Flavor and Juiciness: Moderately juicy.

Zones		Harvest Period
S. Cal. Coastal	1	Jan. - Mid May
S. Cal. Inland	2	Mid Nov. - April
Cal. & Ariz. Deserts	3	Nov. - Jan.
Cal. Central Valleys	4	Mid Nov. - Mid April
N. Cal. Coast. Valleys	5	Dec. - April

'Robertson', a bud sport of the 'Washington' navel, was introduced commercially because it ripened early and was similar to the 'Washington' navel. It never really caught on because the fruit did not meet commercial standards and because of poor tree growth. However, it has been accepted and widely planted in home gardens because of its small, slow growth habit, fair productivity and its ability to bear fruit at a young age. It appears to be more heat-resistant than the 'Washington' navel and has attractive clusters of fruit.

Summernavel

Tree Size: Medium.
Tree Characteristics: More vigorous and spreading than the 'Washington' with an open growth habit. Large leaves and russet bark. Fruit position-- intermediate.
Fruit Size: Similar to 'Washington' but with a thicker rind.
Fruit Color: Deep orange.
Ease of Peeling: Excellent.
Seeds: Usually none.
Holds on Tree: Good.
Flavor and Juiciness: Rich. Moderately juicy.

Zones		Harvest Period
S. Cal. Coastal	1	Not recommended
S. Cal. Inland	2	Mid Feb. - May
Cal. & Ariz. Deserts	3	Not recommended
Cal. Central Valleys	4	Feb. - Mid May
N. Cal. Coast. Valleys	5	Mar. - May

A late ripening eating orange, the 'Summernavel' is a bud sport of the 'Washington' navel discovered in California. In quality, it is similar to other navels but its late-ripening fruit makes it popular. The open growth of the tree makes it easy to espalier. It has large attractive leaves.

Valencia

Tree Size: Large.
Tree Characteristics: Vigorous growing, more so than 'Washington' navel, and globe shaped. Fruit position--intermediate.
Fruit Size: Medium-large.
Fruit Color: Orange.
Ease of Peeling: Poor.
Seeds: Few.
Holds on Tree: Excellent.
Flavor and Juiciness: Flavor sweet but slightly more acid than the 'Washington' navel. Abundant juice.

Zones		Harvest Period
S. Cal. Coastal	1	April - Mid Oct.
S. Cal. Inland	2	Mar. - Mid Aug.
Cal. & Ariz. Deserts	3	Feb. - Mid May
Cal. Central Valleys	4	Mid Feb. - Aug.
N. Cal. Coast. Valleys	5	April - Mid Oct.

'Valencia', originating in either Spain or Portugal, is the most important commercial orange variety in the world. In the United States, it comprises almost half the total orange production. Just as 'Washington' navel is the best eating orange, 'Valencia' is king of the juice oranges.

'Valencia' is adapted to a wide range of conditions and is grown in all citrus areas. Its moderate to high heat requirement affects the ripening date. In desert areas, it may ripen as early as mid-winter. In coastal areas, the fruit may not be edible until middle or late summer. Late-ripening fruit usually comes in after a second spring bloom which results in the tree having two crops of fruit on it at the same time.

'Valencias', being the latest of the commercial oranges, are not generally grown in areas where frost may damage fruit over-wintering on a tree.

The ability of the 'Valencia' to hold its fruit for long periods while it actually improves in quality is unmatched among oranges. Growing both 'Valencias' and navels will provide fresh fruit year-round.

'Valencias' are subject to rind regreening (see page 67), which does not harm edibility of the fruit.

Shamouti

Tree Size: Medium.
Tree Characteristics: Upright, moderately vigorous. Densely foliated with dark green leaves. Few thorns. Fruit position--intermediate. Highly productive in favorable locations.
Fruit Size: Medium-large to large.
Fruit Color: Light orange.
Ease of Peeling: Good.
Seeds: Few or none.
Holds on Tree: Good.
Flavor and Juiciness: Fragrant and pleasantly sweet. Juicy.

Zones		Harvest Period
S. Cal. Coastal	1	Mid Feb. - May
S. Cal. Inland	2	Mid Jan. - Mid April
Cal. & Ariz. Deserts	3	Mid Dec. - Mid Mar.
Cal. Central Valleys	4	Mid Jan. - May
N. Cal. Coast. Valleys	5	Feb. - May

'Shamouti', a popular eating orange in Europe, is an important commercial variety in most areas of the Mediterranean, especially Israel, where it originated. Some fruit reaches supermarkets in the United States in mid-winter.

In the western United States, 'Shamouti' grows well in the same areas as the 'Washington' navel.

Because of its dense foliage, large leaves and large oval-shaped fruit, 'Shamouti' is an attractive and popular tree for home gardens. The fruit does not have a navel.

Hamlin

Tree Size: Medium.
Tree Characteristics: Moderately vigorous, productive. Slightly more hardy than other sweet oranges.
Fruit Size: Medium-small.
Fruit Color: Orange.
Ease of Peeling: Fair.
Seeds: Few or none.
Holds on Tree: Good.
Flavor and Juiciness: Sweet and juicy.

Zones		Harvest Period
S. Cal. Coastal	1	Not recommended
S. Cal. Inland	2	Mid Nov. - Feb.
Cal. & Ariz. Deserts	3	Dec. - Mid Feb.
Cal. Central Valleys	4	Mid Oct. - Mid Feb.
N. Cal. Coast. Valleys	5	Not recommended

'Hamlin' is a major commercial variety world-wide. One of the earliest Arizona Sweets, it actually originated as a Florida seedling. This excellent juice orange is very popular in Arizona, although the fruit may be smaller growing under Arizona conditions than when it is grown elsewhere.

Marrs

Tree Size: Small.
Tree Characteristics: Bears heavy crop of fruit, usually in clusters near the outside of the tree.
Fruit Size: Medium.
Fruit Color: Orange.
Ease of Peeling: Fair.
Seeds: Few to many.
Holds on Tree: Good.
Flavor and Juiciness: Sweet with low acid. Juicy.

Zones		Harvest Period
S. Cal. Coastal	1	Not recommended
S. Cal. Inland	2	Not recommended
Cal. & Ariz. Deserts	3	Mid Oct. - Mid Feb.
Cal. Central Valleys	4	Dec. - Feb.
N. Cal. Coast. Valleys	5	Not recommended

'Marrs' is a naturally dwarf tree producing heavy crops of early fruit. Discovered in Texas, it is believed to be a bud sport of the 'Washington' navel. Primarily grown in Arizona and Texas, it is not commercially grown in California. Earliest of the Arizona Sweets, it is somewhat smaller than most other sweet oranges, possibly because the tree tends to bear fruit at the cost of foliar growth.

Pineapple

Tree Size: Medium.
Tree Characteristics: Moderately vigorous. Few thorns. Productive but less cold tolerant than most citrus.
Fruit Size: Small to medium.
Fruit Color: Orange.
Ease of Peeling: Fair.
Seeds: Few to many.
Holds on Tree: Fair.
Flavor and Juiciness: Rich and sweet. Aromatic. Juicy.

Zones		Harvest Period
S. Cal. Coastal	1	Not recommended
S. Cal. Inland	2	Not recommended
Cal. & Ariz. Deserts	3	Dec. - Mid Feb.
Cal. Central Valleys	4	Mid Dec. - Feb.
N. Cal. Coast. Valleys	5	Not recommended

A popular Florida juice orange in supermarkets, 'Pineapple' is a major commercial variety in Florida where it originated. It is sometimes grown in Arizona, Texas and the Gulf Coast states. The name is derived from the delicate fragrance of the fruit. It is grown infrequently in western home gardens because the fruit is only small to medium in size. It is known as an Arizona Sweet in Arizona.

Diller

Tree Size: Small to medium.
Tree Characteristics: Moderately vigorous, somewhat upright. Productive under desert conditions.
Fruit Size: Fair.
Fruit Color: Orange.
Ease of Peeling: Fair.
Seeds: Few to many.
Holds on Tree: Good.
Flavor and Juiciness: Sweet and juicy.

Zones		Harvest Period
S. Cal. Coastal	1	Not recommended
S. Cal. Inland	2	Not recommended
Cal. & Ariz. Deserts	3	Dec. - Mid Feb.
Cal. Central Valleys	4	Not recommended
N. Cal. Coast. Valleys	5	Not recommended

'Diller' is an early ripening juice variety popular in Arizona where it originated as a seedling. It is an excellent juice orange and well adapted to desert conditions. Sold in Arizona as a Arizona Sweet.

Sour Oranges

The sour or bitter orange family includes many unique ornamental citrus varieties. Their leaves range from the small myrtle leaf of the 'Chinotto', to the rounded leaf of the 'Bouquet de Fleurs', to the elongated, pointed leaf of the 'Seville'. They are small- to medium-sized trees ranging in growth habit from the small, dense, symmetrical rosette of the 'Chinotto', to the rounded, dense, clustered growth of the 'Bouquet de Fleurs', to the upright, stately appearance of the 'Seville'.

There is a difference between "sour" and "bitter" flavors. Sour most specifically refers to fruits with high acid content, such as lemons and limes. Bitterness is caused by essential oils and is a distinctly less pleasant flavor. The English perhaps more appropriately call the 'Seville' and other sour oranges "bitter oranges."

Compared to other orange types, sour oranges often have darker green foliage and more flattened oblate-shaped fruit that tends to be deeper orange in color. Fruit is very sour and aromatic, and is used in drinks, marmalades, liqueurs, orange flower water, perfumes and rind oils. Sour oranges are used to make the liqueurs Curacao and Cointreau. They are also imported into England for manufacturing marmalade. Because the flavor is too strong for most palates, the sour orange has never been very important commercially as a fresh fruit.

The flowers of the sour orange, especially 'Bouquet de Fleurs', are very aromatic. The aroma of leaves and rinds oils is also distinctive. While the fragrance of the sweet orange is pleasant, that of the sour orange blossom is most agreeable and unusual.

Sour orange trees are useful in landscape design. A hardy plant, they are able to withstand conditions too harsh for many other types of citrus. Most varieties make good hedges, and are used extensively as specimen trees. In Arizona and parts of southern California, the 'Seville' is widely used as a street tree.

Sour oranges begin to color in November to December and will hang on the tree for 9 or 10 months. Harvesting periods on the following pages are dates of maturity by commercial grower's standards.

A bowl of 'Chinotto' sour oranges make an attractive centerpiece. The trees are equally handsome in the garden.

Chinotto

Tree Size: Small.
Tree Characteristics: Grows with thornless branchlets into rounded, very symmetrical tree. Leaves are very small, closely spaced and dark green. Dense, compact growth habit.
Fruit Size: Small. Flattened.
Fruit Color: Deep orange. Rind rough and loosely adherent.
Ease of Peeling: Good.
Seeds: Variable from none to many.
Holds on Tree: Excellent.
Flavor and Juiciness: Juicy and sour.

Zones		Harvest Period
S. Cal. Coastal	1	Jan. - Mid Mar.
S. Cal. Inland	2	Dec. - Jan.
Cal. & Ariz. Deserts	3	Nov. - Dec.
Cal. Central Valleys	4	Mid Nov. - Jan.
N. Cal. Coast. Valleys	5	Jan. - Mar.

Small and compact with closely spaced, dense foliage, the 'Chinotto' is a uniquely handsome tree. It produces a profusion of spring blossoms and fruit set in clusters. It originated in Italy where it is prized for making candy. Fruit is also used for jellies and preserves.

The fruit will hold on the tree almost year-round, enhancing its use as an ornamental, container or specimen plant. It may also be trimmed as a hedge or low foundation planting.

Bouquet

Seville

Tree Size: Small.
Tree Characteristics: Spreading tree, grows in thornless branchlets. Dense clustered foliage. Leaves are medium-sized, round, closely spaced.
Fruit Size: Medium. Flattened.
Fruit Color: Deep orange with loosely adherent rind.
Ease of Peeling: Good.
Seeds: Few.
Holds on Tree: Excellent.
Flavor and Juiciness: Juicy and sour.

Zones		Harvest Period
S. Cal. Coastal	1	Jan. - Mar.
S. Cal. Inland	2	Dec. - Jan.
Cal. & Ariz. Deserts	3	Nov. - Dec.
Cal. Central Valleys	4	Dec. - Jan.
N. Cal. Coast. Valleys	5	Feb. - Mar.

The deep orange fruit set against a unique cluster of round, deep green leaves makes the 'Bouquet' (or 'Bouquet de Fleurs') a prized ornamental. Its flowers in massive clusters are probably the most aromatic of all citrus types. It is ideally suited as a patio container plant or in the ground for accent, framing, or as a specimen. It also makes an excellent hedge. As a general rule, the dwarf variety can be used as a shrub wherever the ornamental shrub pittosporum is grown.

Tree Size: Medium.
Tree Characteristics: Vigorous upright growth habit. Thorny. Long, dark green leaves taper to a point.
Fruit Size: Medium. Flatter than a sweet orange.
Fruit Color: Deep orange with rough, loosely adherent rind.
Ease of Peeling: Good.
Seeds: Variable.
Holds on Tree: Excellent.
Flavor and Juiciness: Juicy but very sour.

Zones		Harvest Period
S. Cal. Coastal	1	Jan. - Mar.
S. Cal. Inland	2	Dec. - Mid Feb.
Cal. & Ariz. Deserts	3	Nov. - Dec.
Cal. Central Valleys	4	Dec. - Mid Feb.
N. Cal. Coast. Valleys	5	Jan. - Mar.

A highly ornamental Spanish variety. The largest commercial groves are in Spain. Fruit is shipped to England for making marmalade. Historically, the 'Seville' was used extensively in courtyards and in gardens around mosques. Several Arizona cities now use 'Seville' for roadside planting. An excellent ornamental for the home garden. It is often used in patios, courtyards and alongside drives as individual specimen trees or as background or corner accent.

Blood Oranges

One-third of all the oranges consumed in the Mediterranean region of Europe are blood oranges. Despite their excellent flavor, however, they have not been accepted commercially in the United States.

The name blood orange best describes one aspect of the fruit's uniqueness—its deep red internal color. There are many varieties of blood orange popular around the world, but three are known for their predictable color. They are 'Moro', 'Sanguinelli' and 'Tarocco'. Each has its own distinctive flavor, fruit size and climate requirements.

Unfortunately, the name "blood" does not do justice to the fruit's other exceptional quality—unequalled flavor. In fact, the name "blood orange" probably repels more people than it attracts.

The exact reasons for the development of the red color are not known. The intensity of coloration seems to vary by light, temperature and variety. Most display a red blush on the peel but this does not always indicate internal color. Rind and flesh have different requirements for color development.

The flesh color of blood oranges generally is deepest in hot interior regions. Varieties differ: 'Tarocco' may develop deep color in California interior valleys, but none at all in the desert. Rind coloration will be the most dramatic on fruit that gets some shade, growing on the inside or northern side of the tree. The fact that shade intensifies external blush suggests some interesting experiments with shade cloths or plastics. Rind color adds considerably to the ornamental qualities of the tree.

Blood oranges are called the connoisseur's or gourmet's citrus. Their flavor is distinctive and refreshing. It is often described as a rich orange flavor with overtones of raspberries or strawberries. The fruit is a bit hard to peel but excellent for juicing. Flavor is outstanding in all areas regardless of coloration, but the flavor of the fruit is sweetest in warmer climates.

The fact that their color is unpredictable is probably the main reason why blood oranges have not been accepted commercially in the United States. The juice also has a tendency to turn a muddy color during processing.

For the homemaker, the blood orange, with or without color, offers exotic flavors and ranks high among the tropical fruits.

Blood oranges have a familiar citrus tang with a hint of raspberry. From left: 'Tarocco', 'Moro' and 'Sanguinelli'.

Tarocco

Tree Size: Medium.
Tree Characteristics: Moderately vigorous, somewhat irregular in form. Moderately productive. Fruit held toward inside of tree.
Fruit Size: Medium-large to large.
Fruit Color: Rind, orange, blushed with red at maturity. Flesh, deeply pigmented.
Ease of Peeling: Fair to good.
Seeds: Few.
Holds on Tree: Poor. Tends to deteriorate quickly after maturity.
Flavor and Juiciness: Rich, sweet and distinctive. Juicy.

Zones		Harvest Period
S. Cal. Coastal	1	Mar. - May
S. Cal. Inland	2	Feb. - Apr.
Cal. & Ariz. Deserts	3	Dec. - Feb.
Cal. Central Valleys	4	Mid Jan. - Apr.
N. Cal. Coast. Valleys	5	Mid Feb. - Mid May

A very distinctive-tasting Italian variety, 'Tarocco' is the oldest and most widely grown blood orange in Italy. It is less productive than other blood oranges, but yields bigger fruit with a high juice content. Internal color is unreliable but usually intermediate between 'Sanguinelli' and 'Moro'. The best color develops in interior valleys. Color is poor in the desert, unreliable near the coast. Its open growth habit can cause problems in windy areas but also makes it a suitable tree to espalier. It may be pruned to shape.

Sanguinelli

Tree Size: Small to medium.
Tree Characteristics: Almost thornless. Long, narrow, light green foliage. Very productive. Fruit held on outside of tree.
Fruit Size: Small to medium. Oblong.
Fruit Color: Rind usually blushed deep red. Flesh is orange with red streaks at maturity.
Ease of Peeling: Fair to good.
Seeds: Few to none.
Holds on Tree: Good. Also stores well.
Flavor and Juiciness: Excellent. Juicy. Distinctive flavor.

Zones		Harvest Period
S. Cal. Coastal	1	Mar. - May
S. Cal. Inland	2	Feb. - April
Cal. & Ariz. Deserts	3	Dec. - Feb.
Cal. Central Valleys	4	Mid Jan. - Mid April
N. Cal. Coast. Valleys	5	Mar. - Mid May

'Sanguinelli' is the most popular blood orange in Spain. Relatively dense foliage and an abundance of rich, red, oblong fruit make it a highly attractive addition the home garden. A long holding period extends the harvest time. External red color, unsurpassed by other blood oranges, should develop fairly well, even near the coast.

Moro

Tree Size: Medium.
Tree Characteristics: Vigorous, spreading and round-topped. Very productive. Fruit borne in clusters near outside of tree.
Fruit Size: Medium. Smaller than 'Tarocco' but larger than 'Sanguinelli'.
Fruit Color: Rind, reddish-orange. Flesh, deep red, violet or burgundy.
Ease of Peeling: Fair to good.
Seeds: Few to none.
Holds on Tree: Good.
Flavor and Juiciness: Juicy. Pleasant flavor with distinctive aroma.

Zones		Harvest Period
S. Cal. Coastal	1	Feb. - Mid May
S. Cal. Inland	2	Jan. - Mar.
Cal. & Ariz. Deserts	3	Dec. - Feb.
Cal. Central Valleys	4	Mid Dec. - Mid Apr.
N. Cal. Coast. Valleys	5	Feb. - Apr.

'Moro' is an early-ripening Italian variety. Its distinctive, deep internal color develops under most growing conditions, from the desert to fairly near the coast. Rind color is best in hot interior regions. Fruit is held in clusters near the outside of the tree, making it an attractive ornamental.

Oranges at the Supermarket

Look for relatively heavy oranges with thin skins.

'Valencia' oranges grown in different climates. Left, grown in Florida; right, California. Note differences in rind thickness. Russet coloring of skin is often a sign of juicy, quality fruit.

Once you have found oranges in the produce section of your supermarket, which variety will be your best buy? Which is freshest? Which one is juiciest?

To start with, it helps to know where an orange comes from. Florida and California produce over 90 percent of the oranges that go to market in the United States. While good-natured ribbing goes on between these two citrus industries, both produce excellent oranges. Their climates, however, are quite different and climatic factors control fruit characteristics such as flavor, color, acidity, sweetness, thickness of the rind, ripening time and juice content.

The semitropical climate of Florida has high humidity and warm nights. Rind color, however, develops best with warm days and cool nights, so Florida citrus, with a few exceptions (most notably the 'Dancy' tangerine) are not as brightly colored as those from the drier, subtropical regions of California and Arizona.

Temperature fluctuations between night and day also influence flavor which is judged largely on a sugar-to-acid ration rather than sweetness. The California climate with its warm days and cool nights produces fruit with high sugar and high acid and a richer flavor than the typically sweet fruit from Florida.

Humidity influences juice content and the smooth-ness and thickness of the rind. Florida, with its high humidity, produces a juicier orange than California, with a smoother, thinner rind.

Because Florida oranges are less colorful but juicier than California oranges, they are primarily used for processed juices and orange by-products. The more colorful California ones are usually marketed fresh. Consequently, most of the fresh oranges in the supermarket are probably from California or Arizona.

Timing is another important consideration in citrus selection. Knowing when certain varieties ripen helps judge freshness. It also indicates what should be the best buys at specific times in the year.

Following is a timetable showing the availability of various varieties of oranges. It is, of course, only a guideline. Seasons fluctuate and so do ripening periods. Also, some varieties can hang on the tree months after ripening without deteriorating.

'Valencias' are grown in Florida, California and Arizona. They make the best juice orange and are the most important orange variety in the world. The harvesting period begins in early February and lasts into October. They are found in supermarkets throughout this period.

In Florida, harvesting of 'Valencia' begins in early February and extends into mid July. April and May are peak months. In California and Arizona, 'Valencias' grown in the desert are the first to ripen in early spring. Next come those from the Central Valley and interior regions of California. 'Valencias' from coastal California are the last to ripen.

The ability of 'Valencias' to hang on the tree for four to five months without deteriorating makes them especially prized for the market. They can be picked as the market dictates and this results in a harvest period of about eight months in California. 'Valencias' in cool coastal areas can remain on the tree longer than those in warmer interior regions.

'Washington' navels are the standard eating orange. California is the only area of important commercial production. Harvesting begins in November and lasts into May. Peak supplies are in January, February and March.

'Hamlin', 'Pineapple' and 'Parson Brown' are three varieties widely grown in Florida. They are excellent juice oranges and are major constituents of processed orange juice originating from that state. All three, especially 'Pineapple', occasionally find their way into supermarkets.

Pineapple is probably the most important midseason variety from Florida. It begins ripening in November and continues into March. Most are available in January and February.

California 'Valencia' oranges often go through a natural process called regreening. The fruits are fully ripe and juicy, but the warm temperatures return green chlorophyll to the rind.

'Parson Brown' is an early variety which, if it reaches the market at all, is available in October and November and sometimes December.

The 'Shamouti' or 'Jaffa' orange comes from Israel. It is sometimes available in the United States in winter.

GOOD ORANGES

What other characteristics distinguish a good orange? Packing houses in Florida and California have grading systems on which they base the price of oranges per box. Each state's system is different, but size, sugar-acid ratio, freedom from blemishes, color, juice content and several other fruit qualities enter into the grading. Once the oranges are out of the box and into the bin at the market, you must make your own selection.

A good orange is firm, has smooth skin, is relatively free of blemishes and has no cuts or soft spots. On some oranges you may notice a thin waxy coating. After oranges are picked, they are washed and dipped in a harmless wax to keep them from drying out during shipping and to reduce decay.

Weight is a good indication of juice content. An orange that is heavier than one of similar size will contain more juice.

Size is not always a good way to judge oranges. A small heavy fruit is a better choice than a large light one. A large 'Washington' navel, for example, may just have a thicker rind. The rind is nature's way of protecting the fruit and its thickness is not the sole characteristic on which to base selection.

Color is not usually an indication of quality. Judging an orange by color alone is probably the single most common mistake made by citrus consumers.

As mentioned before, oranges grown in Florida are not as brightly or uniformly colored as those grown in California. Speckling or russet-coloring are common to Florida oranges. In fact, it can be a sign of a very juicy, thin skinned fruit.

Although illegal in California and Arizona, some Florida oranges are colored with a harmless vegetable dye. Such oranges, by law, must be marked "color added."

California 'Valencias' go through a process called *regreening* in late spring and summer. If they remain on the trees during the summer months, the warmth causes chlorophyll to return to the rind, giving it a greenish tinge. The quality of the inner fruit is unchanged, and they are often marketed as "summer greens."

Oranges **67**

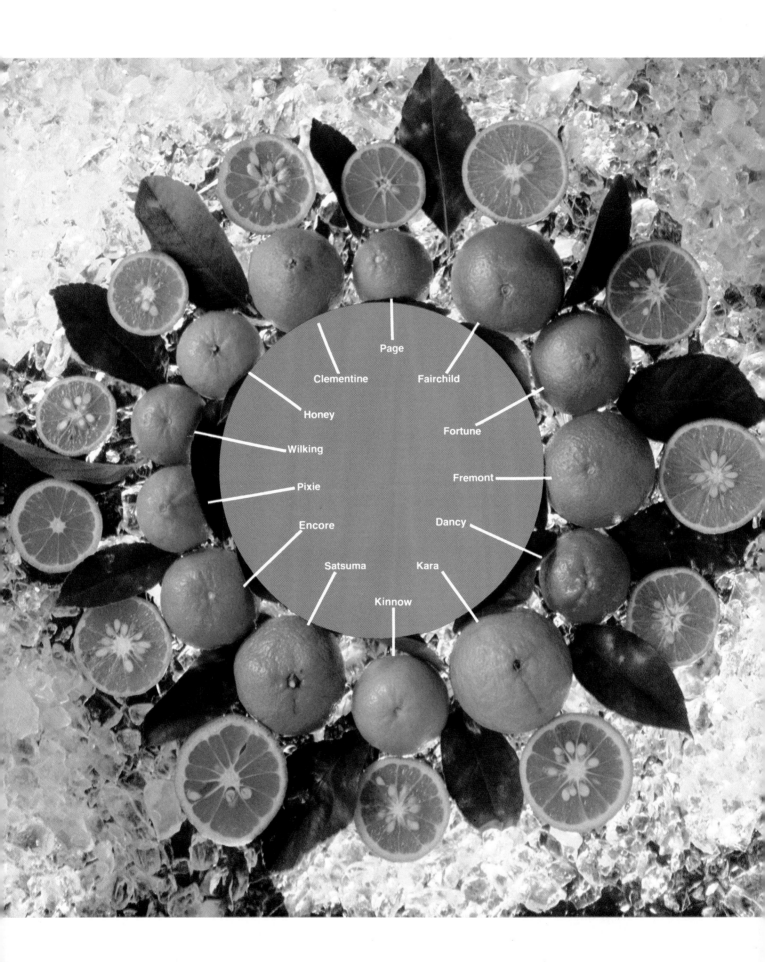

Page

Clementine

Fairchild

Honey

Fortune

Wilking

Pixie

Fremont

Encore

Dancy

Satsuma

Kara

Kinnow

Mandarins

The Mandarins easily make up the largest and most varied group of edible citrus. An entire book could be written on variety, flavor and fruit characteristics and how climate influences their growth. As landscape plants, they provide an array of unique textures and tree forms. There are stately trees such as 'Dancy' and lower weeping forms such 'Wilking' and 'Willowleaf'. To the western fruit grower, mandarins are multipurpose plants that fit into innumerable landscape situations.

Some mandarins are called *tangerines*. **The word tangerine seems to have developed with 'Dancy' which** has a more brightly colored, orange-red peel than most mandarins. Since the introduction of 'Dancy', the varieties with a deeper red coloration have been labeled tangerines, although they are all technically mandarins.

The description of mandarins as "slip-skin" oranges or "kid-glove" fruit refers to the fact it is easy to peel them and separate the fruit into segments. Although mandarins are generally easy to peel, some varieties are harder to peel than others. And if the rind is puffy and extremely easy to peel, the fruit may simply be overripe.

Mandarins usually fall into one of four groups: the Satsumas of Japan, the Mediterranean mandarins of the Mediterranean basin, the King mandarins of Indo-China and the common mandarins, a large group which includes the popular 'Clementine' and 'Dancy'. Mediterranean and 'King' are also the names of specific varieties; Satsuma is used loosely as a variety name, but in reality refers to one of several early-ripening varieties. See page 80.

Most of the newer varieties available today are crosses between two members of these groups and many share the best qualities of each of their parents. Some crosses have resulted in very different and desirable fruit flavors such as the sweetness of 'Honey' mandarin, the rich and aromatic flavor of 'Kinnow', or the sprightly flavor of 'Kara'. Other crosses have lessened undesirable alternate bearing qualities or extended the harvest periods. Thanks to plant breeders, ripening periods begin in November with the popular Satsuma mandarin and extend into June and even July with late varieties such as 'Encore'. With careful varietal selection, as few as three or four mandarin trees can yield fresh fruit for an eight-month period.

Although mandarins have a long and illustrious history originating in the Mediterranean and India, they have been most important commercially in the Orient. Japan still ranks as one of the largest producers of mandarins, primarily Satsumas, but acreage in Florida (the largest U.S. producer) and California is growing rapidly.

Tangors (mandarin orange and sweet orange cross) and tangelos (mandarin orange and grapefruit cross) are often grouped with mandarins because of their mandarin parentage. They are handled separately on pages 102 and 106.

The range of climatic adaptation of the mandarins is the largest among citrus. Trees are hardy and early varieties, such as the Satsumas, are widely grown in colder regions. The fruit, however, may be somewhat cold-sensitive, more so than oranges or grapefruit, thus ruling out late varieties in severe frost areas. Mandarins generally are also very heat tolerant and many varieties are well adapted to desert and hot inland conditions.

The climate in which mandarins are grown has a great deal of influence on fruit quality. High heat in the latter part of the ripening period produces the best flavor. In the desert, the fruit is usually sweeter, juicier and larger than fruit of the same variety grown in cooler climates. Some mandarins, such as the Satsuma and 'Kara', are not adapted to the desert heat and do better under cooler conditions.

Alternate bearing is often a problem with some varieties. Fruit thinning during heavy years may help. Other varieties such as 'Clementine', 'Fairchild' and 'Fortune' require a cross-pollinator for best fruit production. Usually a different variety of mandarin or tangelo which blooms during the same period will suffice. We have listed candidates for cross-pollination under the descriptions of individual varieties.

Mandarins **69**

Clementine

Tree Size: Small to medium.
Tree Characteristics: Grows at moderate rate. Very attractive with weeping form and dense, dark green foliage. Fruit held toward the outside makes this a very showy tree.
Fruit Size: Small to medium.
Fruit Color: Red-orange.
Ease of Peeling: Excellent. Will not become puffy until long after maturity.
Seeds: Varies from few to many.
Holds on Tree: Excellent.
Flavor and Juiciness: Sweet and juicy.

Zones		Harvest Period
S. Cal. Coastal	1	Feb. - Mid Apr.
S. Cal. Inland	2	Jan. - Mar.
Cal. & Ariz. Deserts	3	Dec. - Feb.
Cal. Central Valleys	4	Mid Jan. - Mid Apr.
N. Cal. Coast. Valleys	5	Feb. - Mid Apr.

A North African variety, 'Clementine' is excellent as an early-season mandarin. Sweet juicy fruit which holds on the tree several months past maturity makes the tree an excellent ornamental. It is a major commercial variety in California where the fruit is available in markets from December through April or May. It yields its best fruit in the desert but produces well in cooler climates. It requires a pollinator for best fruit production. 'Dancy' and 'Kinnow' mandarins, 'Orlando' tangelo and 'Valencia' orange are commonly used as pollinators.

Dancy

Tree Size: Medium-large.
Tree Characteristics: Erect growing with a vigorous habit. Few thorns. Fruit held towards outside of tree.
Fruit Size: Small to medium.
Fruit Color: Red-orange.
Ease of Peeling: Excellent, puffy only after maturity.
Seeds: Many.
Holds on Tree: Fair.
Flavor and Juiciness: Rich and sprightly. Moderately juicy.

Zones		Harvest Period
S. Cal. Coastal	1	Mar. - April
S. Cal. Inland	2	Feb. - Mar.
Cal. & Ariz. Deserts	3	Mid Dec. - Mid Feb.
Cal. Central Valleys	4	Jan. - Mar.
N. Cal. Coast. Valleys	5	Mid Feb. - April

Introduced into Florida from Morocco, 'Dancy' is now the leading commercial variety in the United States. The traditional Christmas tangerine, fruit from Florida reaches the markets around the holiday season. Florida, where high heat and humidity yield the sweetest fruit, is the major producer. Desert regions of California and Arizona produce acceptable fruit, but sunburn may be a problem. In cooler areas, fruit is more tart and flavorful but not edible until spring.

The tree needs space in the landscape. It has a tendency to bear in alternate years.

Encore

Honey

Tree Size: Medium.
Tree Characteristics: Moderate growth rate. Upright, spreading habit, rather open. Few or no thorns. Fruit position—intermediate.
Fruit Size: Medium.
Fruit Color: : Orange with darker orange spots.
Ease of Peeling: Excellent.
Seeds: Many.
Holds on Tree: Good.
Flavor and Juiciness: Rich and juicy.

Zones		Harvest Period
S. Cal. Coastal	1	May - July
S. Cal. Inland	2	Mid April - Mid July
Cal. & Ariz. Deserts	3	Mid Mar. - June
Cal. Central Valleys	4	Mid April - Mid July
N. Cal. Coast. Valleys	5	May - Mid Aug.

'Encore' is a cross between 'King' and 'Mediterranean' mandarins. It is a very valuable late-maturing variety which bears colorful speckled fruit of excellent flavor. Recently introduced by the University of California at Riverside. It is a heavy producer on an open tree, but it does have a tendency to bear in alternate years.

Tree Size: Medium-large.
Tree Characteristics: Grows vigorously into spreading tree. Abundance of mid-season fruit makes attractive tree.
Fruit Size: Small.
Fruit Color: Yellow-orange.
Ease of Peeling: Excellent.
Seeds: Many.
Holds on Tree: Good.
Flavor and Juiciness: Rich and very sweet. Juicy.

Zones		Harvest Period
S. Cal. Coastal	1	Feb. - Mid April
S. Cal. Inland	2	Jan. - Mar.
Cal. & Ariz. Deserts	3	Dec. - Feb.
Cal. Central Valleys	4	Mid Jan. - Mid April
N. Cal. Coast. Valleys	5	Feb. - Mid April

Another cross of 'King' and 'Mediterranean' mandarins, this small-fruited but sweet variety is a good choice for the home fruit grower. Generally a heavy producer, it does best in warmer areas, producing fruit that has a superior flavor compared to other sweet mandarin types. This 'Honey' mandarin of California should not be confused with the 'Murcott' mandarin of Florida which is sometimes labeled in supermarkets as 'Honey'.

Fairchild

Tree Size: Small to medium.
Tree Characteristics: Grows at a fairly vigorous rate. Round-headed, dense and nearly thornless. Fruit held toward outside of tree.
Fruit Size: Medium.
Fruit Color: Deep orange.
Ease of Peeling: Fair.
Seeds: Many.
Holds on Tree: Fair to good. Some flavor loss, with time.
Flavor and Juiciness: Rich, sweet and juicy.

Zones		Harvest Period
S. Cal. Coastal	1	Not Recommended
S. Cal. Inland	2	Jan.-Mid Mar.
Cal. & Ariz. Deserts	3	Dec.-Feb.
Cal. Central Valleys	4	Jan.-Feb.
N. Cal. Coast. Valleys	5	Not Recommended

'Fairchild', 'Fortune' and 'Fremont' have 'Clementine' as a common parent. Other parents are: 'Orlando' tangelo of 'Fairchild', 'Dancy' of 'Fortune' and 'Ponkin' of 'Fremont'. They were introduced by the U.S.D.A. Citrus and Date Experiment Station in Indio, California, to produce 'Clementine' quality fruit over a long season in areas of high summer heat. 'Fairchild' produces earlier than 'Clementine' and is the only one of commercial importance. 'Fortune' ripens last; 'Fremont' (cross section shown in photo above) is early bearing but after 'Fairchild'. 'Fremont' and 'Fortune' are subject to wind and sun damage. 'Fairchild' and 'Fortune' require pollination to produce their best fruit. 'Clementine', 'Kinnow', 'Wilking' and 'Fremont' mandarins, and 'Valencia' orange are acceptable pollinators.

Kara

Tree Size: Medium-large.
Tree Characteristics: Moderate growth rate. Open tree similar to, but larger than Satsuma. Large, dark green leaves with a drooping habit. Nearly thornless.
Fruit Size: Medium-large.
Fruit Color: Orange.
Ease of Peeling: Good.
Seeds: Many.
Holds on Tree: Fair. Tends to become puffy shortly after ripening.
Flavor and Juiciness: Rich, sprightly and distinctive. Juicy.

Zones		Harvest Period
S. Cal. Coastal	1	Apr. - May
S. Cal. Inland	2	Mar. - Apr.
Cal. & Ariz. Deserts	3	Not Recommended
Cal. Central Valleys	4	Mar. - Apr.
N. Cal. Coast. Valleys	5	Mid Mar. - Mid June

A California introduction, 'Kara' is a cross between 'King' and Satsuma. It ripens late and is best suited to interior regions. Fruit is good but a bit tart near the coast. Fruit quality is poor in the desert.

Fruit is frost sensitive, limiting its success in colder regions. It tends to bear in alternate years.

Kinnow

Tree Size: Large.
Tree Characteristics: Vigorous growing, upright tree. Foliage is dense and willow-like. Attractive ornamental.
Fruit Size: Medium.
Fruit Color: Yellow-orange.
Ease of Peeling: Fair to good.
Seeds: Many.
Holds on Tree: Good to excellent.
Flavor and Juiciness: Rich, aromatic and distinctive. Very juicy.

Zones		Harvest Period
S. Cal. Coastal	1	Apr.-May
S. Cal. Inland	2	Feb.-Apr.
Cal. & Ariz. Deserts	3	Dec.-Feb.
Cal. Central Valleys	4	Feb.-Apr.
N. Cal. Coast. Valleys	5	Mid Mar.-Mid June

This University of California introduction is increasing in commercial importance. The fruit is delicious and very juicy.

The objective of fruit breeders working on this variety was to combine the fruit quality of 'King' with the tree form of the 'Mediterranean' mandarin. The result is a large, very attractive tree with excellent fruit.

It has a strong tendency to bear in alternate years but is widely adapted to all citrus regions of California and Arizona. It has the willow-like growth habit and fruit with the rich flavor and distinctive aroma of 'Mediterranean' mandarin.

Page

Tree Size: Medium-large.
Tree Characteristics: Moderately vigorous, round-headed tree. Dense foliage, very ornamental. Few or no thorns. Fruit position—intermediate.
Fruit Size: Small to medium.
Fruit Color: Orange-red.
Ease of Peeling: Good.
Seeds: Few to many.
Holds on Tree: Good.
Flavor and Juiciness: Rich and sweet.

Zones		Harvest Period
S. Cal. Coastal	1	Feb. - Mid May
S. Cal. Inland	2	Jan. - Mar.
Cal. & Ariz. Deserts	3	Dec. - Jan.
Cal. Central Valleys	4	Jan. - Mar.
N. Cal. Coast. Valleys	5	Feb. - Mid May

A cross between a 'Minneola' tangelo and 'Clementine' mandarin, 'Page' is commonly called an orange, rather than a mandarin, because of its orange-like appearance. Technically speaking, it should be classified in the tangelo hybrid group since its parentage is three-fourths mandarin and one-fourth grapefruit (pummelo). A recent U.S.D.A. introduction from Florida, it is an exceptionally good tasting, juicy variety that deserves attention. Although the fruit is generally small they are produced in large numbers and are quite ornamental. 'Page' is a great choice for fresh juice lovers; possibly the best of the mandarin hybrids. It needs a pollinator; 'Dancy' or 'Orlando' tangelo, or 'Valencia' orange are adequate.

Mediterranean

Pixie

Tree Size: Small to medium.
Tree Characteristics: Moderately vigorous, broad, spreading tree. Attractive weeping habit with willow-like leaves. Few or no thorns. Tree hardy to cold and resistant to unfavorable conditions. Fruit held towards inside of tree.
Fruit Size: Medium.
Fruit Color: Yellowish orange.
Ease of Peeling: Good.
Seeds: Many.
Holds on Tree: Poor. Becomes puffy and deteriorates quickly after maturity.
Flavor and Juiciness: Sweet, juicy, distinctly aromatic.

Zones		Harvest Period
S. Cal. Coastal	1	Mar. - April
S. Cal. Inland	2	Feb. - Mar.
Cal. & Ariz. Deserts	3	Mid Dec. - Feb.
Cal. Central Valleys	4	Mid Jan. - Mar.
N. Cal. Coast. Valleys	5	Mid Feb. - Mid May

While not of commercial importance in California, 'Mediterranean' mandarin is second only to Satsuma in worldwide importance. Widely grown in both Spain and Italy. High heat produces best fruit but tree is very cold tolerant. Tends to bear in alternate years.

Tree Size: Medium-large.
Tree Characteristics: Vigorous growth with an erect, open habit. Rather open dark green foliage.
Fruit Size: Small to medium.
Fruit Color: Yellowish orange.
Ease of Peeling: Excellent.
Seeds: None.
Holds on Tree: Fair to good. Will have some juice loss and becomes puffy after maturity.
Flavor and Juiciness: Pleasant and mild. Moderately juicy.

Zones		Harvest Period
S. Cal. Coastal	1	Mid Apr. - June
S. Cal. Inland	2	Mar. - Mid June
Cal. & Ariz. Deserts	3	Not Recommended
Cal. Central Valleys	4	Apr. - July
N. Cal. Coast. Valleys	5	Mid Apr. - Mid July

'Pixie' is a late-maturing California introduction best adapted to coastal and intermediate regions. It has a tendency to bear in alternate years. It was hybridized in California through open pollination of 'Kincy' mandarin, a cross between 'King' and 'Dancy'.

Satsuma

Tree Size: Small to medium.
Tree Characteristics: Slow spreading growth. Open, dark green foliage. Tree resistant to unfavorable conditions and is cold hardy.
Fruit Size: Medium.
Fruit Color: Orange.
East of Peeling: Excellent.
Seeds: None.
Holds on Tree: Poor. Fruit becomes puffy and deteriorates quickly after maturity.
Flavor and Juiciness: Mild sweet flavor. Juicy.

Zones		Harvest Period
S. Cal. Coastal	1	Mid Dec. - Mid Apr.
S. Cal. Inland	2	Nov. - Jan.
Cal. & Ariz. Deserts	3	Not Recommended
Cal. Central Valleys	4	Nov. - Jan.
N. Cal. Coast. Valleys	5	Mid Dec. - Mid Apr.

The Satsuma mandarin is an important variety for western growers. The earliest to ripen and very hardy, it can be grown in areas normally too cold for citrus. It is popular in the Sacramento Valley and Sierra foothills of California, and the Gulf Coast of Texas. Fruit ripens before dangerous frosts, and the foliage survives temperatures to about 22° to 18° F (-5° to -8° C). Yields are poor in desert climates.

The name technically refers to a group of early-ripening mandarins, which include the ancient Japanese 'Owari' Satsuma. Most Satsumas available at nurseries are selections of the 'Owari', labeled 'Owari' Satsuma or just Satsuma mandarin. Fruit should be picked when it reaches good color. It does not last on the tree, but stores well refrigerated.

Wilking

Tree Size: Small to medium.
Tree Characteristics: Dense, moderately vigorous. Very ornamental, willow-like foliage. Few or no thorns. Fruit position—intermediate.
Fruit Size: Small to medium.
Fruit Color: Orange.
Ease of Peeling: Good.
Seeds: Many.
Holds on Tree: Good but with some puffiness of skin.
Flavor and Juiciness: Rich and very juicy.

Zones		Harvest Period
S. Cal. Coastal	1	Feb. - Mid May
S. Cal. Inland	2	Jan. - Mid April
Cal. & Ariz. Deserts	3	Dec. - Mid Mar.
Cal. Central Valleys	4	Jan. - Mid Apr.
N. Cal. Coast. Valleys	5	Mid Feb. - May

'Wilking' is a cross between 'Mediterranean' and 'King' mandarins, so has the same parentage as 'Kinnow'. It ranks as one of the most ornamental mandarins with a compact, dense green head and beautiful willow-like leaves. Fruit is deep orange and has excellent flavor. The tree has good cold tolerance. It does have a tendency to alternate bear and when it does bear it produces so many fruit that some limbs can be damaged.

Mandarins at the Supermarket

Mandarin and mandarin hybrids often cause some confusion at the supermarket. As previously explained, some mandarins are often called *tangerines*. To add to this, certain mandarin hybrids, namely tangelos (a cross between a mandarin and a grapefruit) and tangors (a cross between a mandarin and an orange), are mislabeled as oranges, mandarins or tangerines. For this reason, we have grouped buying tips for mandarins, tangelos and tangors in this section rather than in their respective chapters.

Florida produces the most mandarins in the United States with the brightly colored 'Dancy' tangerine selling best. It is the most familiar mandarin to shoppers and reaches the supermarkets just before Christmas and remains there through January. 'Dancy' tangerines from the western states are available from January to April.

The most important early variety in California is 'Clementine', at the market primarily in late November and December.

Satsuma mandarins are exceptionally hardy and grown in colder regions of California. They are early mandarins which reach the market from November into January.

Worldwide, more Satsumas are grown in Japan where they are often processed for canned segments and, to some extent, exported fresh. There are also small acreages of commercially grown Satsumas along the Gulf Coast of the United States.

Satsumas may occasionally retain some green coloration even when completely ripe. Don't let this stop you from experiencing this excellent tasting variety.

The 'Fairchild' is another early-ripening mandarin which is becoming more important commercially in California and Arizona. It is grown primarily in desert regions and reaches the market from late November to February.

Two later varieties from California are finding their way into the supermarkets with greater frequency. They are 'Kara' and 'Kinnow.' Both have excellent flavor and are available between January and April; 'Kara' is slightly earlier than 'Kinnow'.

'Honey' mandarin is a small but delicious variety from California. It is exceptionally sweet and juicy, and available from January to March.

A Florida variety often called 'Honey' is actually the 'Murcott' mandarin, more commonly called 'Murcott' orange. It is larger than the California 'Honey', but is available about the same time or later.

California mandarins, left, have a richer flavor because of their high sugar and high acid content. 'Murcott' (sometimes called 'Honey') mandarin from Florida, right, has a sweeter flavor.

TANGELOS

Three principal varieties of tangelos are available at the supermarket: 'Minneola', 'Orlando' and 'Seminole'. The protruding neck of the 'Minneola' tangelo, with orange-red rind color, is easy to spot in markets from December through April. Florida, California and Arizona each produce 'Minneola' and each has its own flavor and appearance.

'Orlando' tangelo was originally called 'Lake', because it originated in Lake County, Florida. It is often sold as 'Orlando' orange and has an orange-red rind. Available at markets from November to March.

Rind color is deepest orange when days are warm and nights are cool as in California. The 'Murcott' mandarin, right, was grown in Florida. It lacks the deeper color, but will likely be juicier and sweeter than the California grown 'Honey', left.

'Seminole' is probably the least available tangelo. When it does reach the markets, it is late in the season, from February into April and May.

TANGORS

'Temple' orange, as it is called in most stores, is actually a tangor. It is also sometimes called the 'Royal' mandarin. The 'Temple' is a large fruit that really comes into its own under Florida conditions, although those grown in western deserts are quite acceptable. Fruit reaches the supermarkets from late December to March and April.

It is also a good idea to acquaint yourself with the differences between Florida and California citrus as outlined in the section, Oranges at the Supermarket, on page 66.

All mandarins are easy to peel. A little looseness of the rind at the supermarket is not unusual. However, in most varieties, excessive looseness can be a sign of overripeness.

Like any citrus, the heaviest mandarin or mandarin-hybrid will yield the most juice. The skin should be free of bruises and soft spots. Color is often a poor indication of quality. Russet fruit from Florida may be quite good eating.

Lemons

The most obvious characteristic of lemons is that they are acid rather than sweet. Their acid content is at a maximum prior to fruit maturity. For commercial purposes this means that they can be picked by size rather than by ripeness. The earlier they are picked after they obtain sufficient size and juiciness, the more acidic they are and the longer they can be stored. Few lemons that reach the supermarket are actually tree ripe.

The home lemon grower, however, has the opportunity to pick a lemon at the optimum time from his own tree. For best flavor, a lemon should be picked when it becomes fully yellow. Left too long on the tree, they become pithy and lose flavor and acidity.

Because lemons are an acid citrus, they do not need much heat to bring the fruit to full ripeness. On the other hand, lemons are more sensitive to frost than many other citrus types.

Commercially, most lemons are grown in the coastal regions of California. Areas with relatively even temperatures foster the everblooming tendency of certain lemon varieties. In regions characterized by mild winters and moderate summers, such as the San Francisco Bay area, lemon trees may bear fruit throughout the year with only a slight chance they will be harmed by frost. In interior regions, fruit quality is excellent, the fruit reaching maturity in the winter and again in the late summer.

The lemon tree is vigorous, upright, spreading and

'Lisbon' lemon

has an open growth habit. It will attain a large size under favorable conditions. Unlike most other citrus, lemons should be pruned to keep them from getting too tall and rangy. Pruning also results in a more compact and densely foliated tree which adds to its landscape appeal. The leaves are large and light green. The tree tends to flower year-round, making it attractive as an ornamental. Both the flowers and new growth are tinged with purple.

'Eureka' and 'Lisbon' lemons are the major commercial varieties. Ripe fruit is yellow, medium-sized, juicy, highly acid and has very few seeds. The fruit of these two varieties are difficult to tell apart.

Other popular varieties are 'Meyer' and 'Ponderosa,'

'Meyer' lemon

'Eureka' lemon

which are actually hybrids. 'Meyer' is most likely a cross between a lemon and an orange. 'Ponderosa' is presumed to be a cross between a lemon and a citron.

'Meyer' is one exception to the rule that lemons are sensitive to cold. It is as hardy as a sweet orange and probably the most popular landscape citrus in California. It is an attractive plant, a prolific producer and its fruit can be used much as you would use a true lemon.

At one time the widely planted 'Meyer' lemon was carrying a virus infection. Because of the danger to other citrus, its sale was greatly restricted. Recently the University of California introduced virus-free plants under the name 'Improved Meyer' lemon. Make sure you purchase a certified virus-free 'Improved Meyer' lemon. Such trees will bear a yellow tag from the California Department of Food and Agriculture.

'Ponderosa' is known for its large fruit. Lemons are often the size of grapefruit and are produced year-round.

Because of its size, the true lemon tree is usually planted as a specimen or background tree. 'Meyer' and 'Ponderosa' lemon trees are more often planted in patio containers although they make good specimens. Any lemon tree can be made into an espalier. 'Meyer' makes a very attractive hedge.

Eureka

Tree Size: Medium.
Tree Characteristics: Moderately vigorous, somewhat open growth habit. Nearly thornless. Productive with a strong tendency to bear fruit year-round. Frost sensitive. Flowers and new growth tinged with purple.
Fruit Size: Medium.
Fruit Color: Rind, yellow at maturity. Flesh, greenish yellow.
Ease of Peeling: Very difficult.
Seeds: Few to none.
Holds on Tree: Good, but loses acidity; best if picked when ripe.
Flavor and Juiciness: Highly acid and juicy.

Zones		Harvest Period
S. Cal. Coastal	1	Sep. - Aug.
S. Cal. Inland	2*	Oct. - Dec., May - June
Cal. & Ariz. Deserts	3	Sep. - Oct., Mar. - May
Cal. Central Valleys	4*	Nov., May - June
N. Cal. Coast. Valleys	5	Sep. - Aug.

*Some year-round

'Eureka' is a major commercial variety that originated in California. Where frost is not an overwhelming concern, the everbearing tendency of 'Eureka' makes it an excellent choice for the home garden. The ever-present fruit and flowers, in highly visible clusters, make it a handsome tree that can serve as a background, individual specimen or patio container tree.

The fruit ripens nearly year-round in coastal areas with main harvest in late winter to early spring. In warmer inland areas there are two crop cycles.

Lisbon

Ponderosa

Tree Size: Large, up to 30 feet on a vigorous rootstock.
Tree Characteristics: Vigorous growth, densely foliated, upright. Thornier than 'Eureka'. Fruit held on inside of tree. Most productive and cold-hardy of the true lemons. Flowers and new growth tinged with purple.

Fruit Size: Medium.
Fruit Color: Rind, yellow at maturity. Flesh, greenish yellow.
Ease of Peeling: Very difficult.
Seeds: Few to none.
Holds on Tree: Good, but looses acidity; best if picked when ripe.
Flavor and Juiciness: Highly acid. Juicy.

Zones		Harvest Period
S. Cal. Coastal	1	Sep. - Aug.
S. Cal. Inland	2*	Oct. - Dec., May - July
Cal. & Ariz. Deserts	3	Sep. - Oct., April - June
Cal. Central Valleys	4*	Oct. - Dec., May - July
N. Cal. Coast. Valleys	5	Sep. - Aug.

*Some year-round

'Lisbon' is a major commercial variety world-wide that originated in Portugal. It is more tolerant of heat, cold, wind and neglect than other true lemons, and preferred in desert and inland locations. Dense foliage and heavy fruit-set add year-round color.

Tree Size: Small to medium.
Tree Characteristics: Vigorous, roundheaded, thorny and productive. Large leaves. Flowers and new growth tinged with purple. Sensitive to frost.

Fruit Size: Medium-large to large.
Fruit Color: Rind, yellow. Flesh, pale green.
Ease of Peeling: Very difficult.
Seeds: Many.
Holds on Tree: Very good.
Flavor and Juiciness: Acid and juicy.

Zones		Harvest Period
S. Cal. Coastal	1	Sep. - Mid Aug.
S. Cal. Inland	2	Sep. - Aug.
Cal. & Ariz. Deserts	3	Sep. - Aug.
Cal. Central Valleys	4	Sep. - Aug.
N. Cal. Coast. Valleys	5	Sep. - Aug.

A natural hybrid of lemon and citron, this variety was first grown in the United States in Maryland, but it is probably of Italian origin. The fruit of 'Ponderosa' can be substituted for a true lemon. The rind is quite thick due, probably, to its citron heritage. It produces lemons the size of grapefruits year-round and is an unusual ornamental. It can be grown as a container or specimen plant; hedge or espalier.

Improved Meyer

Tree Size: Small to medium.
Tree Characteristics: Moderately vigorous and spreading. Nearly thornless. Hardy and productive. More or less everblooming. Fruit held near inside of tree.
Fruit Size: Medium.
Fruit Color: Rind, yellowish orange at maturity. Flesh, light orange-yellow.
Ease of Peeling: Very difficult.
Seeds: Few to many.
Holds on Tree: Good. Sweetness increases on tree.
Flavor and Juiciness: Slightly sweeter than true lemon but still with excellent, acidic, lemon-like flavor. Very juicy.

Zones		Harvest Period
S. Cal. Coastal	1	Sep. - Aug.
S. Cal. Inland	2	Sep. - Aug.
Cal. & Ariz. Deserts	3*	Sep. - Oct., May - July
Cal. Central Valleys	4	Sep. - Aug.
N. Cal. Coast. Valleys	5	Sep. - Aug.

* Some year-round

Not long after this 'Meyer' lemon hybrid was imported from China, it became the most popular citrus for the home garden, particularly in California. Usually grown as a rooted cutting, it will often bear fruit the first year.

The tree is hardy and very ornamental; the fruit is thin-skinned and juicy. Although ideal for containers, it also makes an excellent hedge.

Lemons at the Supermarket

Of all citrus, lemons have the widest range of uses. This may be hard to understand when you think of the response of a person first biting into a lemon. But as any good chef will tell you, lemons have a unique ability to enhance or accentuate the flavor of other foods. The oils responsible for this quality are quite complex and are used widely in the food industry. For more information on the use of lemons, see page 166.

Virtually all commercial lemons in North America originate in California and Arizona. Most come from coastal regions in the southern part of California and the largest percentage of these come from Ventura County. The cool coastal climate accentuates the everblooming character of lemons and they are harvested virtually year-round. This, and the fact that lemons store well, result in year-round availability at the supermarket. Harvest periods for lemons grown in warmer, more inland areas are concentrated in late fall and winter.

'Lisbon' and 'Eureka' are two varieties available at the supermarket. Although there are some tree differences, the fruits are difficult to distinguish.

A good lemon will have a smooth, thin rind and feel heavy for its size. Lightweight fruit with a thick, coarse rind usually will have less juice. The color should be bright yellow, and there should be no bruises or soft spots. Lemons with green coloration will be more acid and less flavorful.

When you have a choice at the supermarket, look for a lemon with a smooth, thin rind, and one that appears heavy for its size. It will usually have more juice than the lightweight fruit with a coarse rind.

Grapefruit

Three principal varieties of grapefruit are grown in the United States: the white-fleshed 'Marsh', the very similar red-fleshed 'Redblush' ('Ruby'), and the white-fleshed 'Duncan'. 'Duncan' is grown mainly in Florida.

Grapefruit need extremely high heat before they reach maximum eating quality. In the long, hot growing season of the California low desert and in Arizona and Texas, they reach their prime in about 12 to 14 months. Grapefruit grown in cooler regions closer to the coast, take as long as 18 months to ripen. The trees will commonly have two crops on them at one time. Generally, this fruit will be more tart and have a thicker rind than fruit grown in the desert.

Because of the differing ripening periods of the two growing areas, grapefruit grown near the coast are commercially profitable when desert fruit is not avail-able. It also means that many tart grapefruit reach the supermarket, but that some kind of grapefruit are available almost 12 months of the year. Many people prefer the tart fruit.

Grapefruit trees are ornamental enough to be grown for that reason alone. They are large, the foliage is a glossy dark green and the fruit is borne in attractive clusters near the outside of the tree.

There are two things you can do to increase the sweetness of grapefruit grown in colder climates. First, you can plant in the warmest possible location. This might be in an area with a southern exposure or in front of a wall that will reflect the heat of the sun.

Second, you can let the fruit hang on the tree for as long as possible, up to 18 months. The fruit will be sweeter if held past maturity. Eventually, however, if held too long, the flavor will become insipid. If you let the fruit hang on the tree for long periods, it will usually result in two crops growing on the tree at once, greatly adding to its ornamental value.

Grapefruit has many uses but is most often associated with breakfast. Left is 'Marsh Seedless'; right is 'Redblush'.

Marsh Seedless

Tree Size: Large, tall.
Tree Characteristics: Vigorous growth into productive, spreading tree. Glossy, deep green foliage. Fruit position intermediate, but may be very visible. Many are borne in clusters.
Fruit Size: Large.
Fruit Color: Rind, pale to light yellow. Flesh, buff-colored.
Ease of Peeling: Fair.
Seeds: Few or none.
Holds on Tree: Excellent.
Flavor and Juiciness: Good flavor and very juicy.

Zones		Harvest Period
S. Cal. Coastal	1	May - Nov.
S. Cal. Inland	2	Feb. - Sep.
Cal. & Ariz. Deserts	3	Jan. - Mid May
Cal. Central Valleys	4	Mid Jan. - June
N. Cal. Coast. Valleys	5	April - Mid Aug.

'Marsh' (or 'Marsh Seedless') originated in Florida, and is the standard white fleshed grapefruit. It is also the parent of the most popular red-fleshed grapefruit, 'Redblush'. Except for the obvious difference in flesh and rind color of the fruit, 'Redblush' rind has a pink blush with proper heat. 'Marsh' and 'Redblush' are almost identical trees.

Grapefruit compote.

Like all grapefruit, 'Marsh' will reach its peak quality only with high, prolonged heat. Desert conditions are ideal. Other climates yield fruit which is less sweet and ripens later. However, in some areas of the San Joaquin Valley, fruit quality approaches that of desert-grown fruit.

Both 'Marsh' and 'Redblush' trees are large, up to 25 feet tall on standard rootstock, or 8 feet and taller on dwarfing rootstock.

Redblush

Tree Size: Large, tall.
Tree Characteristics: Same as 'Marsh' but pink-blushed fruit can add a different ornamental quality.
Fruit Size: Large.
Fruit Color: Rind develops crimson blush. Flesh is light pink under favorable conditions.
Ease of Peeling: Fair.
Seeds: Few to none.
Holds on Tree: Excellent.
Flavor and Juiciness: Good flavor and very juicy.

Zones		Harvest Period
S. Cal. Coastal	1	May - Nov.
S. Cal. Inland	2	Feb. - Sep.
Cal. & Ariz. Deserts	3	Jan. - Mid May
Cal. Central Valleys	4	Mid Jan. - June
N. Cal. Coast. Valleys	5	April - Mid Aug.

'Redblush' is a leading pigmented variety and is identical to 'Marsh' in almost all aspects except flesh and rind color. It is the standard red-fleshed grapefruit but only develops internal color with proper heat. Near the coast, it is difficult for fruit to develop pigmentation. For additional comments, see 'Marsh', page 89.

Grapefruit at the Supermarket

The quality of grapefruit depends to a great extent on where it is grown. Because grapefruit requires high heat for a sweet flavor, many people feel the best fruit comes only from the warmer areas. In Florida, that is the Indian River Valley. The best western fruit comes from the California and Arizona deserts and Texas. Much of the Texas grapefruit is sold through mail order outlets, as are some of the Florida grapefruit.

Florida and Texas grapefruit reach the market in September and last into May. October through April are peak months. Western grapefruit from desert areas is available from January to June. Cool areas of California produce summer grapefruit which lasts into September. This means grapefruit is available at the supermarket year-round.

Florida, California and Texas account for about 75 percent of the grapefruit grown in the world. Florida and Texas produce nearly 90 percent of the grapefruit in the United States.

There are three principal commercial grapefruit varieties.

White-fleshed 'Duncan' grows primarily in Florida. Despite its seediness, many people consider it best because of its excellent flavor.

'Marsh Seedless' is a white-fleshed variety that grows in all commercial citrus states. 'Marsh' is the parent of the red-fleshed grapefruit which is available under many names such as 'Thompson', 'Ruby' and 'Redblush'. Each is a distinct strain. 'Redblush' is the most popular red-fleshed variety and has the deepest internal color. Although the red external blush and red interior make 'Redblush' a consumer favorite, its flavor is indistinguishable from 'Marsh'.

A fourth variety that occasionally reaches supermarkets is new from Texas, named 'Star Ruby'. Its main attraction is its deep red coloration of both peel and flesh.

As with other citrus, the heaviest grapefruit yields the most juice. Fruit should be firm and smooth-skinned and either round or flattened. Avoid fruit with rough, puffy rinds.

Rind color is not a good indication of interior quality. Russet fruit is often the best flavored. Furthermore, some grapefruit go through a regreening process similar to that which occurs in 'Valencia' oranges (see page 67). It does not affect quality.

Here are a few of the many grapefruit brand names you'll find at the market. All are from the warmest areas of southern California, Arizona, Texas and Florida.

Limes

Like lemons, limes are an acid fruit. They are sensitive to cold and should be grown in frost-protected places. The two species of acid limes are classified as *large-fruited* and *small-fruited* limes. In the western United States, the small-fruited limes are represented by the 'Mexican' variety which is commonly referred to as the *bartender's* lime. In Florida, the 'Mexican' lime is known as the 'Key' lime.

The large-fruited limes are represented by the variety 'Bearss'. It is believed to be identical or very similar to the 'Tahitian' or 'Persian' lime of Florida.

Surprisingly, there are more differences than similarities between the 'Mexican' and 'Bearss' lime. The differences between the two trees are very pronounced. The 'Mexican' is finer stemmed, thornier, much less vigorous, and has smaller leaves of a distinctively pale color. Compared to 'Bearss', 'Mexican' is more sensitive to cold and requires more heat to develop good sized fruit.

The fruit differences are also quite pronounced. 'Bearss' lime is larger, virtually seedless, less aromatic and less flavorful than the 'Mexican' lime.

'Bearss' lime is grown commercially in California. 'Mexican' lime grows best in the warmer, more frost-free semitropical climates found in parts of Mexico and Florida.

On dwarf rootstock 'Mexican' grows into a round, spreading shrub suitable for containers or useful as a foundation plant. By planting them in a sunny, wind-protected area, you may be able to create a semi-tropical-like microclimate that will enable you to enjoy the fruit of this highly aromatic variety.

In contrast, 'Bearss' lime is medium-sized, vigorous and broad-spreading. In the home landscape, 'Bearss' can be used as an individual specimen. Because it is hardy, 'Bearss' is often grown successfully in areas reserved for lemons.

The bartender's lime, left, grows in the semitropical regions of Mexico and Florida. The 'Bearss' lime, right, is grown throughout California.

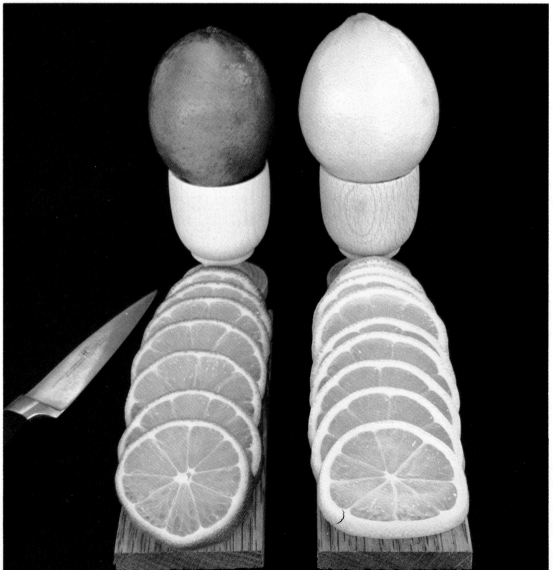

Mexican

Tree Size: Small.
Tree Characteristics: Grows at a moderate rate into twiggy tree. Leaves small but dense. Many small thorns. Fruit position intermediate but closer to outside.
Fruit Size: Small.
Fruit Color: Rind, greenish to yellow-orange at maturity. Flesh, straw yellow.
Ease of Peeling: Very difficult.
Seeds: Few to many.
Holds on Tree: Poor. Turns yellow at maturity, then drops.
Flavor and Juiciness: Highly acid and aromatic. Juicy.

Zones		Harvest Period
S. Cal. Coastal	1	July - Dec.
S. Cal. Inland	2	July - Dec.
Cal. & Ariz. Deserts	3	July - Oct.
Cal. Central Valleys	4	July - Nov.
N. Cal. Coast. Valleys	5	Aug. - Dec.

Highly aromatic, the 'Mexican' lime is often called the bartender's lime. Few areas of the western United States have the right climate for their growth. By planting dwarfs, however, and exploiting micro-climates, you may be able to enjoy the distinctive qualities of this fruit. They require a combination of long, warm summers and frost-free winters. Container planting in a sheltered location or in front of a warm, south-facing wall may duplicate these conditions.

Commercially, the 'Mexican' lime is picked when green even though they are not fully ripe until they turn yellow. There is very little flavor difference between the green and yellow fruit, but they should not be picked when they are too small.

Bearss

Tree Size: Medium.
Tree Characteristics: Vigorous with dense green foliage, roundheaded. Some thorns but less than 'Mexican' lime. Fruit position intermediate but closer to outside of tree; sometimes very visible.
Fruit Size: Similar to a small lemon.
Fruit Color: Rind, pale yellow at maturity. Flesh, pale greenish yellow.
Ease of Peeling: Very difficult.
Seeds: Few to none.
Holds on Tree: Fair to good. May be some deterioration.
Flavor and Juiciness: True acid lime flavor. Very juicy.

Zones		Harvest Period
S. Cal. Coastal	1*	Aug. - Mar.
S. Cal. Inland	2*	Feb. - Mar., June - Aug.
Cal. & Ariz. Deserts	3	June - Sep.
Cal. Central Valleys	4*	June - Mar.
N. Cal. Coast. Valleys	5*	Aug. - Mar.

* Some year-round

'Bearss' is a Tahitian variety introduced in Porterville, California. Like the lemon, 'Bearss' does not require much heat to reach ripeness. Besides being as hardy as a lemon, it is a beautiful ornamental producing colorful, fragrant blossoms and very juicy fruit. The fruit

has a yellowish green color when fully ripe but can be used at an earlier green stage. The rind is thin and shiny. 'Bearss' is without doubt the most valuable lime for western gardeners; it has an extended fruit season, and it is highly productive. It also makes an excellent tree for a front entrance or for a container.

Rangpur

Tree Size: Medium sized, spreading and drooping.
Tree Characteristics: Vigorous and very productive. Slender twigs and comparatively few or small thorns. Leaves dull green. New growth tinged with purple. Very cold hardy.
Fruit Size: Small to medium. Round and loose skinned.
Fruit Color: Rind, reddish orange. Flesh, orange.
Ease of Peeling: Very good.
Seeds: Many.
Holds on Tree: Very good.
Flavor and Juiciness: Tender, juicy and very acid.

Zones		Harvest Period
S. Cal. Coastal	1	Nov. - Dec. in all areas
S. Cal. Inland	2	and very everblooming.
Cal. & Ariz. Deserts	3	Nov. - Dec. in all areas.
Cal. Central Valleys	4	
N. Cal. Coast. Valleys	5	

'Rangpur' is not a true lime. While most of the characteristics resemble a mandarin it is often grouped with limes because of its acid flavor. It makes an acceptable lime substitute.

It is a prolific producer of reddish orange fruit that hang on the tree an extremely long time. The overlap of bloom and colorful fruit, and the purple tinge of its new growth highlight its ornamental qualities. It is very attractive as a container plant.

The popular indoor plant, 'Otaheite' orange, is an acidless dwarf form of the 'Rangpur'. The fruit has an insipidly sweet flavor.

Limes at the Supermarket

Two types of limes are sold in American supermarkets. The large fruited type is from the Tahitian group and is called 'Persian' lime in Florida and 'Bearss' lime in California. The second type is the familiar aromatic 'Mexican' lime, often referred to as the bartender's lime.

'Persian' or 'Bearss' limes are grown commercially, primarily in Florida and to a lesser extent in California. 'Mexican' limes prefer tropical climates and reach supermarkets from Mexico and other South American countries. A few are grown in southern Florida.

Limes are available nearly year-round with the biggest shipments from Florida occurring during June to August. California limes are on the market from August to February. Limes from Mexico are available nearly year-round.

Limes are marketed green which is when their acid content is at its peak. At full maturity they turn yellow.

'Persian' or 'Bearss' limes should have a firm rind with a glossy, medium green color. Avoid yellow fruit because they will have less acid.

'Mexican' limes are usually dark green, but some yellow is all right as long as they are firm.

Avoid any lime with a dull, dry skin or with brown or purple coloration. Make sure they are firm and heavy. The heaviest lime yields the most juice.

Center is a display of good quality 'Mexican' limes. The deep green color is typical.

These three fruit are all 'Mexican' limes. Avoid any like the one at left. It will be fairly dry inside. A yellow 'Mexican' lime is okay if firm. The dark green, firm and heavy lime at right is at peak quality.

Pummelos

It is natural to describe the pummelo in relation to the grapefruit because they are so closely related. The size of pummelo fruit is well reflected by its botanical name *Citrus grandis*. Its fruit, the largest among citrus, is as big as a large grapefruit or bigger. It is generally round to pear-shaped, with a thick skin, firm flesh and a lower juice content than grapefruit.

Because of the firm flesh and the scanty juice, you do not eat pummelos in the same manner as you do grapefruit. Instead, you peel the fruit, segment it and shell the edible pulp vesicles out of their membranes which you can do without rupturing the vesicle walls. The shelled segments are eaten with or without sugar.

In their natural habitat, pummelos are a varied group. Besides differing in size and shape, their fruit also differs in flavor. Some pummelos are very sour; some are sugar-acid; some non-acid (like a sweet orange). Although the common (sugar-acid) pummelos have pale to lemon-yellow flesh, there are also pigmented varieties with flesh ranging from pink to deep red. Many are very attractive with excellent flavor. Some are very seedy, others are almost seedless. There are also double pummelos, analogous to the double-fruited navel oranges.

The common and pigmented varieties, primarily of Siamese origin, have relatively thin rinds and a mildly acid flavor. They are most important commercially. In the Far East where pummelos have long been favored for eating as a fresh fruit, some of the most noteworthy varieties are the 'Kao Pan', 'Siamese Pink' and 'Kao Phaung' of Thailand, the 'Mato Butan' of Taiwan, the 'Bonpeiyn' of Japan and the 'Moanlua' of Tahiti.

In the United States the pummelo is considered a collector's item, primarily because of its giant-sized fruit.

The trees vary in size, shape and growth habit. Some are small and round-topped; others are equal in size to the grapefruit and just as vigorous. Some are drooping while others are large, open and wide-spreading; some are nearly thornless, others have large prominent thorns.

Pummelos are similar to grapefruit but larger and with a thicker rind. 'Chandler', left, is pigmented and most popular here, but 'Reinking', right, is similar in flavor. The edible pulp vesicles of these and other varieties are favorites throughout the Far East.

The twigs are generally thick and hairy for citrus. The leaves are distinct with their broad, winged leaf stems. Flowers are large and woody in texture.

Closely akin to grapefruit, pummelos are about the same or slightly more sensitive to frost. In general, they need somewhat less heat to ripen the fruit.

Recent breeding work at the University of California's Citrus Experimental Station in Riverside has resulted in some interesting fruit types. The most available is 'Chandler' which was created by crossing 'Siamese Pink' with 'Siamese Sweet' pummelo. It is intermediate in flavor between the two, being somewhat sugar-acid and pigmented like its pollen parent, 'Siamese Pink'. A small quantity of 'Chandler' is grown commercially in southern California primarily to serve the Oriental communities in San Francisco and other parts of the state. Pummelos are in great demand for the Chinese New Year.

'Reinking' was developed at the U.S. Department of Agriculture's Citrus and Date Experiment Station at Indio, California. It performs well in California under the right climatic conditions. Its flavor is also classified as sugar-acid.

A number of hybrids have been developed by crossing the low-acid (sweet) pummelo with other citrus types. A sweet pummelo-grapefruit and a sweet pummelo-mandarin cross have yielded superb fruit. The grapefruit hybrid is large, very juicy and tastes like a sweet grapefruit. The mandarin hybrid is extremely juicy and sweet. In both cases, their flavor is halfway between that of both parents, offering two new taste experiences. They also ripen with much less heat them a grapefruit. Fruit is ripe in December in Riverside, California, where grapefruit will not ripen until March or April.

It is hoped these distinct new citrus will find their way to the public, both as commercial fruit in the market and as nursery plants for the home garden.

How to Peel a Pummelo

To fully enjoy the flavor and nutrition of the pummelo fruit, you should follow these simple steps.

Knife makes peeling 'Chandler' pummelo easier.

First, simply slice off either end.

Here the inner portion of the membrane or pith is cut away.

This attractive salad uses complimentary colors and flavors of sliced navel oranges and 'Chandler' pummelo.

With top sliced away, the rest of the peel is easily pulled free.

Peel is removed and fruit is opened like an orange.

Outer membrane can be peeled away by hand or knife.

Now, fruit is sectioned easily.

Chandler

Tree Size: Medium.
Tree Characteristics: Vigorous. Open growth habit. Leaves are large and broadly winged. Flowers are large and woody.
Fruit Size: Medium and round, with medium-thick rind.
Fruit Color: Rind, yellow. Flesh, pink.
Ease of Peeling: Moderately adherent for a pummelo.
Seeds: Many.
Holds on Tree: Will hold fairly well with slight loss of juice. Sometimes drops after maturity.
Flavor and Juiciness: Flesh firm but tender. Moderately juicy. Good sugar-acid flavor.

Zones		Harvest Period
S. Cal. Coastal	1	Mid April - Mid Aug.
S. Cal. Inland	2	Mid Jan. - Mid May
Cal. & Ariz. Deserts	3	Dec. - Mid April
Cal. Central Valleys	4	Mid Jan. - May
N. Cal. Coast. Valleys	5	Mid April - Mid Aug.

If you don't live in one of the very warm areas of California, Arizona or Texas, you will have to create a microclimate to bring the fruit to full maturity. In the right location it is a heavy producer of pink-fleshed fruit slightly larger than a grapefruit. 'Chandler' is a hybrid developed by the University of California at Riverside and released in 1961. A beautiful, open tree, it serves well as an individual specimen or as a background.

Reinking

Tree Size: Large.
Tree Characteristics: Vigorous tree with denser foliage than 'Chandler'. Fruit is positioned toward the inside.
Fruit Size: Large. Smooth rind.
Fruit Color: Rind, yellow. Flesh, pale yellow.
Ease of Peeling: Moderately adherent for a pummelo.
Seeds: Many.
Holds on Tree: Fruit tends to fall after maturity.
Flavor and Juiciness: Moderately juicy. Flesh firm and tender with good sugar-acid flavor.

Zones		Harvest Period
S. Cal. Coastal	1	Mid April - Mid Aug.
S. Cal. Inland	2	Mid Jan. - Mid May
Cal. & Ariz. Deserts	3	Dec. - Mid April
Cal. Central Valleys	4	Mid Jan. - Mid April
N. Cal. Coast. Valleys	5	Mid April - Mid Aug.

'Reinking' produces abundant white-fleshed fruit, equal in flavor to 'Chandler', but not as popular, probably due to its lack of internal pigmentation. To grow and produce, it requires as much heat as 'Chandler'. It is recommended as an individual specimen or background tree.

Tangelos

Tangelos are hybrids, the result of a cross between a mandarin and a grapefruit or a mandarin and a pummelo. In many instances, the fruit and form of the tree will strongly resemble one of its parents. For instance, 'Minneola' tangelo is a hybrid between 'Dancy' mandarin and the popular Florida grapefruit 'Duncan'. The fruit has a unique flavor similar to a mandarin. 'Sampson' tangelo is a cross between a grapefruit and 'Dancy' mandarin, and has a flavor suggestive of a grapefruit. Other varieties have characteristics that appear to be intermediate between both parents.

In general, tangelos will be most productive with cross-pollination from a mandarin or a tangor. Tangelo varieties do not cross well with one another.

You can peel the fruit of a tangelo rather easily but not as easily as a mandarin.

With the exception of 'Sampson', which isn't recommended near the coast, tangelos are widely adapted. They will, of course, have a more tart flavor in cooler areas. Their hardiness falls between that of a grapefruit and an orange; 'Orlando' is most hardy.

Tangelos are flavorful new citrus hybrids becoming increasingly available in supermarkets. The result of a cross between a grapefruit and a mandarin, they have the best features of both.

Minneola

Tree Size: Medium-large.
Tree Characteristics: Vigorous. Roundheaded tree. Deep green leaves, pointed and large. Many fruit held toward the outside of the tree.
Fruit Size: Large, usually with a prominent neck.
Fruit Color: Orange-red.
Ease of Peeling: Good.
Seeds: Few to many.
Holds on Tree: Good.
Flavor and Juiciness: Rich and tart, unique flavor. Juicy and aromatic.

Zones		Harvest Period
S. Cal. Coastal	1	Mid Feb. - Apr.
S. Cal. Inland	2	Feb. - Apr.
Cal. & Ariz. Deserts	3	Jan. - Feb.
Cal. Central Valleys	4	Feb. - Apr.
N. Cal. Coast. Valleys	5	Mid Feb. - Mid May

'Minneola' tangelo foreground, with its parents, the 'Dancy' mandarin, left rear, and 'Duncan' grapefruit.

'Minneola' is the most important commercial tangelo in California. The bright orange-red fruits with the conspicuous neck are becoming quite common in local supermarkets. However, much of this fruit comes from Florida, the major U.S. producer.

'Minneola' is probably the best tangelo for home gardens. The attractive color of the fruit and its position near the outside of the tree make it a good orna-mental. For best fruit production, use a cross-polli-nator. 'Dancy', 'Clementine' and 'Kinnow' mandarins, or 'Temple' tangor are proven pollinators.

Orlando

Tree Size: Medium-large.
Tree Characteristics: Similar to 'Minneola' but slightly less vigorous. Distinctively cupped, deep green leaves. Fruit held toward outside of tree.
Fruit Size: Medium to large. Rather flattened.
Fruit Color: Orange.
Ease of Peeling: Fair to difficult.
Seeds: Many.
Holds on Tree: Fair.
Flavor and Juiciness: Mildly sweet, closer to a mandarin and very juicy.

Zones		Harvest Period
S. Cal. Coastal	1	Not Recommended
S. Cal. Inland	2	Jan. - Mid Apr.
Cal. & Ariz. Deserts	3	Mid Dec. - Feb.
N. Cal. Valleys	4	Mid Jan. - Mid Apr.
N. Cal. Inland	5	Mid Feb. - Mid May

'Orlando', a cross between a 'Duncan' grapefruit and a 'Dancy' mandarin, is slightly hardier than 'Minneola' and especially adapted to hot desert regions. Fruit is a bit smaller than 'Minneola' but very juicy and matures earlier. It is not recommended for cool coastal areas. For best fruit production, it requires a pollinator; see 'Minneola' for candidates.

Sampson

Seminole

Tree Size: Large.
Tree Characteristics: Vigorous growth into tree resembling a grapefruit. Glossy, cupped leaves. Fruit held near outside of tree.
Fruit Size: Medium.
Fruit Color: Orange-yellow.
Ease of Peeling: Fair to poor. Thin skinned.
Seeds: Many.
Holds on Tree: Good.
Flavor and Juiciness: Acid, similar to a grapefruit.

Zones		Harvest Period
S. Cal. Coastal	1	Not Recommended
S. Cal. Inland	2	Feb. - Mid May
Cal. & Ariz. Deserts	3	Mid Jan. - Mar.
Cal. Central Valleys	4	Feb. - Mid May
N. Cal. Coast. Valleys	5	Not Recommended

'Sampson', a cross between a grapefruit and a 'Dancy' mandarin, is of limited importance and limited availability even though the golden fruit and glossy foliage make it a good ornamental.

Tree Size: Medium to large.
Tree Characteristics: Vigorous, resembling 'Minneola'. Fruit held toward outside of tree.
Fruit Size: Medium-large.
Fruit Color: Red-orange.
Ease of Peeling: Good.
Seeds: Many.
Holds on Tree: Fair.
Flavor and Juiciness: Sprightly and acid. Juicy.

Zones		Harvest Period
S. Cal. Coastal	1	Mid Mar. - Mid May
S. Cal. Inland	2	Mid Feb. - Mid Apr.
Cal. & Ariz. Deserts	3	Feb.
Cal. Central Valleys	4	Mid Feb. - Mid Apr.
N. Cal. Coast. Valleys	5	Mid Mar. - Mid May

'Seminole', a cross between a 'Duncan' grapefruit and a 'Dancy' mandarin, differs from other tangelos; it does not require cross-pollination. Fruit grown in the West, however, is too tart for most people.

Tangors

A tangor is a cross between a mandarin and an orange. Its fruit is like that of a mandarin, but its trees are less hardy. Fruit normally called mandarins may actually be natural mandarin-orange hybrids. For example, 'Clementine' and 'King' mandarins may really be natural tangors.

The tangor's range is limited. The two most commonly planted varieties are 'Temple' and 'Dweet'. The 'Temple' is primarily grown in Florida. In the West, it reaches its full potential only in California, Arizona and Texas deserts and hot inland areas. 'Dweet' can be grown in coastal and inland areas but sunburns badly in desert regions.

Tangors at the Supermarket are discussed in the mandarin section, page 80.

The 'Temple' tangor shown here in orchard row is a natural hybrid grown primarily in Florida. There it is often called the Temple orange.

The tangor, front, is the result of a cross between a sweet orange, right rear, and a mandarin, left rear.

Temple

Tree Size: Small to medium.
Tree Characteristics: Moderately vigorous, spreading, rather shrubby plant. Many thorns. Fruit position, intermediate.
Fruit Size: Medium-large.
Fruit Color: Red-orange.
Ease of Peeling: Good.
Seeds: Many.
Holds on Tree: Fair.
Flavor and Juiciness: Rich and spicy. Moderately juicy.

Zones		Harvest Period
S. Cal. Coastal	1	Not adaptable
S. Cal. Inland	2	Not adaptable
Cal. & Ariz. Deserts	3	Jan. - Feb.
Cal. Central Valleys	4	Not adaptable
N. Cal. Coast. Valleys	5	Not adaptable

The 'Temple' orange, as it is commonly called, had its origin as a Florida seedling. It is an important commercial variety in Florida where it reaches its peak quality. In the West, 'Temple' reaches its distinctive rich and spicy flavor only in the hot deserts. In other areas it becomes unattractive, tart and dry. The tree is more sensitive to cold than either mandarins or oranges.

Dweet

Tree Size: Medium
Tree Characteristics: Moderately vigorous, rather open growth habit. Fruit held near outside of the tree.
Fruit Size: Medium-large.
Fruit Color: Red-orange.
Ease of Peeling: Fair to poor.
Seeds: Many.
Holds on Tree: Poor. Becomes puffy and drops soon after ripening.
Flavor and Juiciness: Rich flavor. Very juicy.

Zones		Harvest Period
S. Cal. Coastal	1	April - May
S. Cal. Inland	2	Mar. - May
Cal. & Ariz. Deserts	3	Not adaptable
Cal. Central Valleys	4	Mar. - May
N. Cal. Coast. Valleys	5	April - June

The 'Dweet' is a rich-flavored fruit that grows well in the interior and coastal regions of California. It is a cross between a 'Mediterranean' sweet orange and a 'Dancy' mandarin. The reddish orange fruit is egg-shaped and has a slight neck. It is sweet to slightly tart in flavor and very juicy. Fruit is held near the end of the branches and subject to sun and wind damage in desert regions.

Kumquats

The kumquat is a small, orange-colored fruit. The rind is thick, tender and sweet; the flesh is moderately acid. Thus the whole fruit, rind and all, is edible—a unique feature among citrus.

The fruit of the kumquat is often used for table decorations, particularly during the Christmas season. They may be clipped with a short section of the stem with several leaves attached.

In China and Japan, kumquat or "Chinan," translates as golden orange or golden bean. It is frequently the custom in China to place fruit-bearing bonsai plants on the table during dinner so guests can pick and eat the fruit between courses.

The most popular kumquat variety in the United States is 'Nagami'. The fruit is small and oval with very few seeds. In some climates, 'Nagami' is slightly tart and has been relegated for use in preserves or for decorations.

The oval kumquat is so popular few people realize that there are also round kumquats. The best of the round kumquats is the 'Meiwa' variety. It is the best for eating fresh. Compared to the 'Nagami', the 'Meiwa' is larger, sweeter and has a more tender rind. Because of these qualities, this Chinese variety is a good choice for the home garden.

A dwarf kumquat will grow to about 3 to 6 feet in height. A standard kumquat grows to about twice that height or taller. Trees tend to be dwarfed, but in rare instances a standard-sized tree has reached a height of 25 feet.

Because the tree is naturally small and symmetrical, with attractive dark green leaves and brightly colored fruit, kumquats are prized as ornamental plants. Dwarf varieties make excellent container plants for terraces and patios. They can also be used as foundation plants or hedges. Standards make beautiful specimen trees.

Kumquat trees need relatively high heat to grow well. As a result they usually remain dormant in fall, winter and spring. During summer months they grow, bloom and set fruit. The amount of heat and humidity determines the size and juiciness of the fruit. In Florida, where it is both hot and humid, kumquats are generally larger and juicier than those grown in California. In California and Arizona deserts, the high heat produces large fruit. But in the coastal regions of California, cool moist air produces small, juicy fruit.

The extended period of dormancy for kumquats probably contributes to their being the most cold-hardy of all citrus species.

The more oblong 'Nagami' kumquat is most familiar, but the round 'Meiwa' is usually preferred for eating fresh.

Meiwa

Tree Size: Small to medium.
Tree Characteristics: Virtually indistinguishable from 'Nagami'. See next page.
Fruit Size: Small but larger than 'Nagami'. Short oblong to round.
Fruit Color: Bright orange.
Ease of Peeling: Not needed.
Seeds: Few.
Holds on Tree: Excellent.
Flavor and Juiciness: Rind, medium thick, tender and sweet. Flesh, orange, slightly juicy and moderately acid.

Zones		Harvest Period
S. Cal. Coastal	1	Jan. - Mid Mar.
S. Cal. Inland	2	Mid Nov. - Feb.
Cal. & Ariz. Deserts	3	Nov. - Dec.
Cal. Central Valleys	4	Mid Nov. - Feb.
N. Cal. Coast. Valleys	5	Dec. - Mid Mar.

The 'Meiwa' is a very popular variety in China and Japan. In addition to being used in preserving, making candied fruit and marmalades, 'Meiwa' is the best kumquat for eating fresh, rind and all. A prized ornamental, it is highly productive when grown in a warm, sunny location.

Kumquats are among the best citrus ornamentals and have many landscape uses.

Nagami

Tree Size: Small to medium.
Tree Characteristics: Evergreen. Fine stemmed, few or no thorns, dense foliage. Small, dark green, pointed leaves.
Fruit Size: Small, oval. Rind medium thick and sweet.
Fruit Color: Flesh, orange. Rind, light orange.
Ease of Peeling: Not needed.
Holds on Tree: Excellent. Will hold for months without loss of flavor.
Flavor and Juiciness: Rind mildly sweet. Juice scanty. Moderately acid.

Zones		Harvest Period
S. Cal. Coastal	1	Dec. - Mid Mar.
S. Cal. Inland	2	Mid Nov. - Feb.
Cal. & Ariz. Deserts	3	Nov. - Dec.
Cal. Central Valleys	4	Mid Nov. - Feb.
N. Cal. Coast. Valleys	5	Dec. - Mid Mar.

The most popular United States variety, 'Nagami' is grown in Florida and California as a Christmas season fruit. It is used primarily for preserving, in making syrup, for candying and making marmalade. The symmetrical tree with a profusion of brightly colored fruit is an attractive ornamental, prized as a container plant.

Ideal container plants, kumquats are frequently used as patio specimens.

Kumquat Hybrids

The limequat, orangequat, citrangequat and calamondin make up an interesting group of ornamental citrus hybrids. The first three originated as part of a breeding program sponsored by the United States Department of Agriculture. The calamondin is a natural hybrid, generally considered a cross between a kumquat and a sour mandarin.

The purpose of the breeding program was to take advantage of the extreme cold-hardiness of the kumquat. The limequat is one of the most useful of these hybrids, a 'Mexican' lime crossed with kumquat, because it is a good lime substitute and is much more cold hardy than its lime parent. Named limequat varieties include the 'Eustis', 'Lakeland' and 'Tavares'.

Another useful member of this family is the orangequat. The 'Nippon' orangequat, a cross between the 'Meiwa' kumquat and the Satsuma mandarin, produces tasty kumquat-like fruit.

The citrangequat group are trigeneric (the result of interbleeding members of three genera) hybrids, a cross of citrange and kumquat. The citrange tree is itself a cross between the sweet orange and the trifolate orange. The fruit is very sour. It is used primarily in marmalades and other foods calling for a sour-acid flavor. The 'Sinton' and 'Macciaroli' are two good ornamental varieties of citrange.

The calamondin, of Chinese origin, is grown extensively in the Philippines under the name *calamonding*. It is used there to flavor food and drinks as lemons or limes are used in the United States. The calamondin is probably the citrus best suited for use as an indoor ornamental plant. It produces good fruit both indoors or outdoors. The variegated calamondin is another very ornamental calamondin; both fruit and foliage are variegated.

All the kumquat hybrids are prolific fruit producers. Fruit of the calamondin and citrangequat will hang on the tree for a long time. The profusion of fruit and extended holding time overlapped by the succeeding season's bloom greatly enhance the ornamental qualities of these hybrids. The limequat and orangequat will also hold on the tree several months before they lose their flavor and juice content.

The leaves of kumquat hybrids vary somewhat in character but their leaves resemble the kumquat leaf in size and shape. The calamondin, orangequat, citrangequat and the 'Tavares' limequat are dense, compact shrubs or small trees. All of these hybrids are good choices for container or patio plants.

Limequats, shown in the foreground, were produced by crossing a kumquat with a 'Mexican' lime, shown at rear. Limequat plants are ornamental and their fruits have flavor and juice like a lime.

Limequat

Tree Size: Dwarf variety grows as a 3- to 4-foot shrub.
Tree Characteristics: Angular open growth habit. Leaves small and round-tipped. Highly productive, very cold-hardy compared to lime parent.
Fruit Size: Oval, slightly smaller than 'Mexican' lime.
Fruit Color: Yellow.
Ease of Peeling: Thin skinned, difficult to peel.
Seeds: Few to many.
Holds on Tree: Good.
Flavor and Juiciness: Flavor, juiciness and aroma of lime.

Zones		Harvest Period
S. Cal. Coastal	1	Dec. - July
S. Cal. Inland	2	Mid Nov. -July
Cal. & Ariz. Deserts	3	Mid Nov. - Mid July
Cal. Central Valleys	4	Mid Nov. - July
N. Cal. Coast. Valleys	5	Dec. - July

'Eustis' is most popular in the western United States. It and 'Lakeland' are similar in size, form and composition, but it is slightly orange when fully ripe. The 'Tavares' tree has a more compact growth habit than the other two. All three tend to be everblooming with the main crop setting in winter. The fruit does not require high heat to ripen. All limequats make excellent container and specimen plants or foundation plantings.

Orangequat

Tree Size: Small.
Tree Characteristics: Dense, spreading tree. Dark green foliage.
Fruit Size: Small, but larger than a kumquat. Round to oval; has neck.
Fruit Color: Red-orange.
Ease of Peeling: Fair. Rind thick, relatively spongy.
Seeds: Few.
Holds on Tree: Excellent.
Flavor and Juiciness: Sweet rind. Pulp, slightly acid and juicy.

Zones		Harvest Period
S. Cal. Coastal	1	Dec. - Sep.
S. Cal. Inland	2	Dec. - July
Cal. & Ariz. Deserts	3	Nov. - Mar.
Cal. Central Valleys	4	Nov. - Aug.
N. Cal. Coast. Valleys	5	Dec. - Sep.

A small, compact tree, it is a prolific producer of mild-flavored fruit during the Christmas season. Very cold-hardy, it is excellent for areas marginal for kumquats; fruit takes much less heat to ripen. Makes an excellent container, specimen or foundation plant. Fruit may be eaten like a kumquat or used in marmalades.

Calamondin

Tree Size: Upright and columnar growth habit.
Tree Characteristics: Very shapely, almost thornless. Finely textured branchlets. Leaves small, broadly oval and mandarin-like. Highly productive and very cold-hardy.
Fruit Size: Small, generally spherical.
Fruit Color: Rind, red-orange. Flesh, orange.
Ease of Peeling: Thin, smooth rind easily separated at maturity.
Seeds: Few.
Holds on Tree: Excellent, almost year-round.
Flavor and Juiciness: Pulp is tender, acid, juicy. Juice sweetened with sugar makes a very palatable drink.

Zones		Harvest Period
S. Cal. Coastal	1	Dec. - Sep.
S. Cal. Inland	2	Dec. - July
Cal. & Ariz. Deserts	3	Nov. - Mar.
Cal. Central Valleys	4	Nov. - Aug.
N. Cal. Coast. Valleys	5	Dec. - Sep.

Calamondin is often placed in the loose-skinned mandarin group. We have placed it with the kumquat hybrids to emphasize its kumquat-like qualities.

It is a spectacular ornamental, bearing hundreds of small bright orange fruit. Useful as a container or specimen plant, to accent a corner or as a hedge.

Citrangequat

Tree Size: Upright, compact.
Tree Characteristics: Moderately vigorous, nearly thornless and highly productive. Cold-hardy.
Fruit Size: Small, oval, and often has neck.
Fruit Color: Reddish orange.
Ease of Peeling: Fairly tight-skinned.
Seeds: Few to none.
Holds on Tree: Excellent.
Flavor and Juiciness: Sharply acid.

Zones		Harvest Period
S. Cal. Coastal	1	Dec. - Sep.
S. Cal. Inland	2	Dec. - July
Cal. & Ariz. Deserts	3	Nov. - Mar.
Cal. Central Valleys	4	Nov. - Aug.
N. Cal. Coast. Valleys	5	Dec. - Sep.

This attractive, symmetrically rounded ornamental is very productive and hardy enough to survive in areas too cold for kumquats. The 'Macciaroli', an Arizona-Texas variety, is similar to 'Sinton', but the fruit is more acid and the tree has a greater tendency towards trifoliation (having three leaflets). It is probably the preferred ornamental because its showy clusters bloom almost year-round and have a gardenia-like fragrance.

Citron

No citrus book would be complete without mention of the citron. The citron was the first citrus to reach the western world and be brought under cultivation. It probably originated in northeastern India. About 300 B.C. it spread to Media and Persia where it became known to the Hebrews, Greeks and Romans. At that time it was called the Persian or Median apple. Most biblical scholars agree that the Hadar or "goodly fruit" of the Bible is the citron. The 'Etrog' citron is the fruit used by the Jewish people at their Feast of the Tabernacles.

Like the lemon and the lime, the citron is a sour fruit. Unlike the lemon and lime, the flesh or pulp of the citron is scanty, firm and lacking in juice.

The fruit is large and oblong. The yellow rind is very thick, fleshy and tightly adherent. It can be smooth, but it is often bumpy. The rind oil is pleasantly aromatic. While the tree is not especially pretty, the citron probably was cultivated because the fragrance of the fruit is delicate, penetrating and long lasting. It is used as a perfumer for rooms and clothing. It has also been used as a moth repellent.

As early as the second century the citron was used in certain epicurean dishes. The white inner part of the peel was used in salads and small pieces of the peel, mixed with herbs, vinegar, oil and spices, were served with fish. Because the peel is high in vitamin B, a method of candying the peel was developed and that is the most important use of the citron today.

The citron grows as a somewhat straggly, small tree or shrub and is sensitive to frost. Its leaves are large, leathery and somewhat oval. The flowers are large and purple-tinged and are produced throughout the year.

The most popular variety is the 'Etrog' citron. In the Orient, the 'Buddha's Hand' citron is widely grown as an ornamental.

In India and areas where the citron is indigenous, naturally occurring hybrids abound. The 'Ponderosa' lemon (probably from Italy) is an example of a citron and lemon hybrid.

The first citrus known to western explorers, the citron is too seldom grown today. However, the white inner part of the peel is candied and available in the spice section of most markets.

Etrog

Tree Size: Moderately vigorous.
Tree Characteristics: Large leathery leaves. New growth and flower buds are tinged purple. Very cold-sensitive. Short life cycle compared to other citrus.
Fruit Size: Large, oblong. Deeply ridged surface.
Fruit Color: Yellow.
Ease of Peeling: Very difficult.
Seeds: Many.
Holds on Tree: Very good.
Flavor and Juiciness: Rind very aromatic. Fruit has little pulp and is quite juiceless but acid.

Zones		Harvest Period
S. Cal. Coastal	1	
S. Cal. Inland	2	Harvest concentrated
Cal. & Ariz. Deserts	3	in fall but is generally
Cal. Central Valleys	4	everbearing in all areas.
N. Cal. Coast. Valleys	5	

'Etrog' citron is the ceremonial fruit of the Jewish Feast of the Tabernacles. Its fragrance is pervasive and can grace a room for weeks. The peel can be candied or used fresh in salads.

As a shrub, it can be used as a foundation or container plant. Everbearing, with an open growth habit. Large, leathery foliage and big fruit.

Fingered Citron

Tree Size: Grows as an open shrub. Moderately vigorous.
Tree Characteristics: Large, rounded, leathery leaves. Very frost-sensitive.
Fruit Size: Medium-sized fruit split into a number of finger-like sections.
Fruit Color: Yellow.
Ease of Peeling: Very difficult.
Seeds: Many.
Holds on Tree: Very good.
Flavor and Juiciness: Fruit very aromatic but has only a small amount of pulp.

Zones		Harvest Period
S. Cal. Coastal	1	
S. Cal. Inland	2	Harvest concentrated
Cal. & Ariz. Deserts	3	in fall but is generally
Cal. Central Valleys	4	everbearing in all areas.
N. Cal. Coast. Valleys	5	

The 'Fingered' citron or 'Buddha's Hand' with its unique finger-shaped fruit is well known and highly esteemed for its fragrance in China and Japan. As an everblooming shrub, it is also valued in the Orient as an ornamental.

The food reserves manufactured by the leaves are also stored in the leaves, twigs and branches. Pruning foliage reduces fruit yield.

Most of the feeder roots of a mature citrus tree are in the top 2 feet of the soil. They extend out twice as far as the drip line. Water and fertilizer application should include this area beyond the drip line.

4

Citrus blossoms are borne on new growth. Only a small percentage of the spring blossoms actually become mature fruit.

Fruit position in the canopy depends on the variety. This influences the ornamental quality of the tree as well as the susceptibility of the fruit to sunburn.

The bark of the citrus is thin and very susceptible to sunburn. It should always be protected if exposed to sun.

The bud or graft union is the point at which the rootstock and fruiting variety join.

Growing Citrus Indoors, Outdoors, Everywhere

Citrus culture varies from garden to garden and climate to climate, but general truths emerge. This is the story of how citrus grows wherever they are grown.

Citrus trees are evergreen, not deciduous. They drop their leaves continuously, a few at a time, rather than all at once in fall. This distinction is important. Evergreen fruit tree growth and culture are very different from deciduous fruit trees. For an understanding of some of these differences, let's follow citrus tree growth from seed to maturity.

SEEDS
Citrus seeds are not particularly difficult to germinate. You might want to experiment with a few left over from lunchtime. Unlike deciduous fruit tree seeds, citrus seeds need no chilling before they will germinate. In fact, the seeds will often begin to germinate inside an overripe fruit. Carefully placed in soil, they will usually grow.

The best temperatures for citrus seed germination are between 80 and 90°F (27 to 32°C).

Nucellar embryony is an important characteristic of citrus seed reproduction. Simply put, this means that some citrus seeds contain more than one embryo capable of developing into a seedling.

Nucellar embryony also means that one seed may produce more than one seedling. One of the seedlings may be a hybrid, the result of normal sexual reproduction. The others will be identical to the parent. They are called *nucellar seedlings*.

The different kinds of citrus vary in the frequency nucellar embryony occurs. For instance, 'Dancy' and

'Chandler' pummelo

'Marsh Seedless' grapefruit

'Eureka' lemon

'Washington' navel orange

'Bouquet' sour orange

'Sanguinelli' blood orange

'Owari' Satsuma mandarin

'Meyer' lemon

'Dancy' mandarin

'Mediterranean' mandarin

'Nippon' orangequat

Variegated calamondin

'Nagami' kumquat

'Eustis' limequat

'Chinotto' sour orang

Leaf characteristics are variable. These 15 kinds of citrus are distinguishable by leaves, but using leaf characteristics to distinguish between varieties of one kind of citrus is usually impossible. Leaf density, color and size are greatly modified by climate.

'Kara' mandarins produce 100 percent nucellar seedlings. 'Eureka' lemons produce only 30 percent.

Because of their predictable growth habit, nucellar seedlings are used primarily for rootstocks onto which chosen varieties are grafted. This joining of rootstock and scion is the best way to get a vigorous, productive tree in the shortest time. However, a seed from the citrus fruit you enjoy may grow into a tree, but will probably not produce good fruit. Also, it will be many years before you will know either way.

SEEDLINGS

Seedling citrus usually develop long taproots. In nursery production the taproot is most often cut during transplanting. This helps promote lateral root formation.

Certain dwarf citrus nurseries use another method of propagation to speed production and better adapt the plants to container growing. Called *twig grafting,* a small branch of the desired variety is grafted to an unrooted cutting of the desired rootstock. In a greenhouse the two varieties grow together at the same time that the rootstock cutting is growing roots. The roots are generally well-branched and more suited to container culture.

The point at which the scion joins the rootstock is called the *bud* or *graft union.* Deformation at the point of union gives an indication of scion-rootstock compatibility. Poorly compatible unions often result in short-lived trees. Overgrowth at the bud or graft union, however, is not necessarily undesirable. The natural restrictions caused by a less than perfect union may induce earlier, more abundant fruiting and desirable plant dwarfing.

ROOTS

Citrus roots are shallow compared to other trees with similar top growth. Feeder roots grow fairly close to the surface. Most are in the top two feet of soil. Also, citrus feeder roots extend well beyond the "drip line," that imaginary line directly below the canopy edge. Once the tree is established in open soil, the watering basin should extend past the edge of the canopy to encourage maximum root growth. Fertilization should also take place within this basin.

LEAVES

Citrus leaves play a role more important than leaves of deciduous fruit trees. They manufacture the plant's food and store any excess. Extra food is also stored in

The 'Eureka' lemon provides a continuous harvest in areas of coastal California. Flowers, developing fruit and mature fruit appear on this tree at one time.

the twigs and branches. Very little is stored in the roots, where deciduous fruit trees store their excess food. The maximum amount of stored food is present in the leaves in late February or early March, just before spring growth. This greatly affects the timing of pruning. Any pruning or leaf loss at this time would of course be at the expense of the fruit harvest.

Cold Tolerance—Citrus does not become dormant in the same sense as a deciduous tree. But when temperatures go much below 54°F (12°C), there is no growth. The tree becomes, in a way, dormant. This is important because citrus trees growing slowly or not at all are more cold-hardy than if they are actively growing.

Some parts of southern California and other warm winter areas may have a problem with this characteristic. Cold snaps, especially if they occur after a few days of unseasonably warm weather, are particularly damaging. The trees will be effectively less hardy than in colder areas such as the San Joaquin Valley.

The fertilization program is also affected by this dormant-hardiness phenomenon. Excess nitrogen in late summer and fall increases the chance of frost damage. It promotes tender new growth at exactly the wrong time. Late winter (January and February) is the preferred time for nitrogen feeding.

FLOWERS AND FRUIT

Citrus flowers form on new growth in late winter and early spring, though some may appear in fall or after periods of drought. Post-drought flowers rarely result in good-eating fruit. Lemons bloom continuously, a habit promoted by a cool coastal climate, where trees will produce several crops each year.

There are always many more flowers produced than eventual fruit. Depending on the climate, especially heat and available moisture, often less than one percent set fruit. The rest fall off.

The first high temperatures of late spring or early summer result in what is called June drop. At this time a large percentage of the immature fruit will fall off. The amount of drop depends on the weather and the tolerances of the variety.

Some citrus varieties, such as 'Washington' navel, are particularly sensitive to high heat at the time of fruit set. Therefore, they are unsuited for many desert areas.

During the flowering and fruit setting period, an adequate water supply is important. Moisture stress at this time can cause virtually all the potential fruit to drop.

Pollination is accomplished by insects and occasionally by the wind, but is not necessary for fruit production with most commercial varieties. Some are self-pollinating. Others, such as 'Fairchild' and 'Clementine' mandarin, require cross pollination from another variety.

Fruit—Alternate bearing—producing fruit every other year—is a characteristic of some varieties of citrus. Exact reasons why are not well understood, but the "no-fruit" year usually follows a year of heavy production. Fruit thinning in the "heavy" year of the cycle helps even out production.

Where the fruit is borne on the tree is important. Some varieties make fruit well within the leaf canopy; others are exposed outside the canopy. Fruit at the outside (and particularly on the southwest side of the tree) will ripen faster, but those on the inside are protected and are less frequently damaged by sunburn.

Seediness—Why do some varieties of citrus have no seeds at all while other varieties have many? Or, why does one fruit have a few seeds while another fruit of the same variety has several? The answers tend to be complicated and depend on the variety.

One reason 'Washington' navel and other navel oranges lack seeds is their lack of viable pollen. Occasionally, cross pollination from another variety produces a few random seeds. More importantly, seedlessness is due to *parthenocarpy,* the ability of a fruit to develop without pollination. Satsuma mandarins do produce some pollen, but will usually be seedless unless there is cross pollination. Other varieties are seedless due to parthenocarpy.

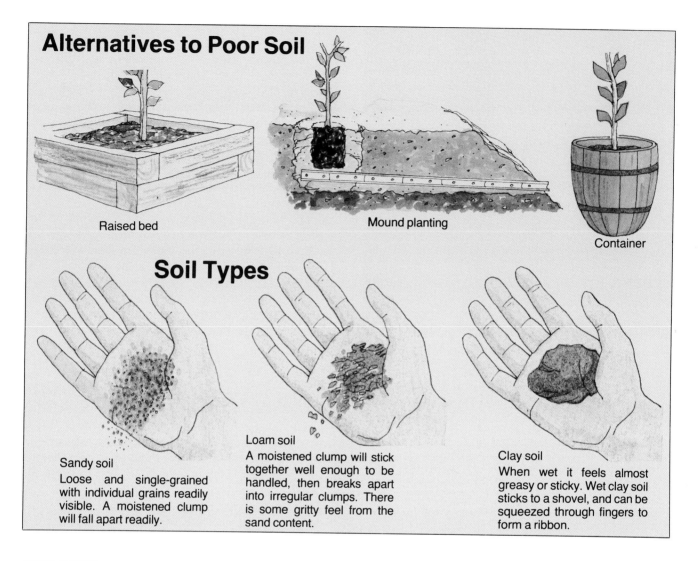

Alternatives to Poor Soil

Raised bed

Mound planting

Container

Soil Types

Sandy soil
Loose and single-grained with individual grains readily visible. A moistened clump will fall apart readily.

Loam soil
A moistened clump will stick together well enough to be handled, then breaks apart into irregular clumps. There is some gritty feel from the sand content.

Clay soil
When wet it feels almost greasy or sticky. Wet clay soil sticks to a shovel, and can be squeezed through fingers to form a ribbon.

SOIL TYPE

Soils are generally classified as clay, loam or sandy, depending on the relative proportions of large and small mineral particles of which they are composed.

Clay particles are the smallest. Predominately clay soils are easily overwatered. Too much water drives out air and encourages fungus disease.

Sand particles are largest. Sandy soils will have plenty of room for air, but moisture and nutrients dissipate quickly. Drought is the common problem of growing citrus in sandy soils.

If your soil is extreme in either of these directions—clay or sandy—it would be wise to use an organic material, such as peat moss, compost, sawdust or ground bark, to amend it. But generally, citrus trees are adaptable and are grown successfully in a wide range of soil types. Commercial groves are located in every type of soil from river bottom sand to heavy black adobe. As a gardener, your challenge is to allow for any unfavorable characteristics your soil may have. Water and

fertilize sandy soils more frequently. Water clay soils less frequently. In the home garden with a limited number of trees, amending the soil will make your gardening easier and far more pleasurable. Excessive amounts of organic matter, especially under the root ball, will allow the tree to settle below the proper soil line increasing disease susceptibility.

Some soils may create more problems than is practical to contend with. For example, salty soil is a common problem in desert regions. Under such conditions, citrus will grow poorly and perhaps die. Salt-burned leaves become brown at their outer edges first. If salty or otherwise unfit soil is a problem, the best solution is to plant in raised beds, containers or mounds.

Container soils—A custom soil mix is the most straightforward way to solve native soil problems. Containers or raised beds filled with a blend of organic material and mineral aggregates (perlite or sand) allow for good drainage and retain nutrients and moisture at the same time.

Dwarf tree   Standard tree

Dwarf or standard, the young tree's structure should be well balanced and capable of supporting full fruit crops. The leaves should have good color, and the bark should be clean and free of defects. Dwarf trees should begin branching three to nine inches above the graft depending on the type. Standard trees should begin branching 24 to 30 inches above the graft.

Left, a familiar sight in nurseries: rows of dwarf citrus in 5 gallon cans.

For only a few trees, a prepared mix is most economical. "Potting soil" comes under many brand names such as Jiffy-Mix, Ready-Earth or Super-Soil. Many gardeners modify the purchased product with perlite or vermiculite to suit their own needs.

Most prepared soil mixes have an initial supply of nutrients added. But all container soils need fairly frequent replenishment of nutrients. That is why it is important to begin a regular fertilization program in at least one month. Check the product label.

TREE SELECTION

Success in growing attractive and productive citrus depends largely on the health and vigor of the purchased tree. For the most part, you can determine this beforehand by examination at the nursery.

Leaves should be healthy, green in color, large, uniform and free from pest damage. Bark should appear bright and clean. The bud union should be smooth. A straight trunk with uninterrupted growth is desirable.

The graft union of a dwarf citrus tree should be about six inches above the soil level. A standard-size tree should be grafted 8 to 12 inches above the soil level.

The tree's framework branches should provide a structure capable of supporting good crops of fruit. Dwarf citrus should begin to branch about three to nine inches above the graft union. A standard-size tree should be encouraged to branch at about 24 to 30 inches above the graft union.

Avoid any trees with extensive circling or matted roots along the inside of its container. This is usually a sign that the tree has been in the container too long. Balled-and-burlapped (B&B) trees often suffer from broken rootballs. Such trees should be avoided.

It is a temptation to buy citrus already loaded with fruit or flowers. But the fact is, a younger plant without fruit or flowers will have an easier time establishing itself after planting than one that has to spend energy ripening fruit. Also, a mature tree has probably been in the container too long.

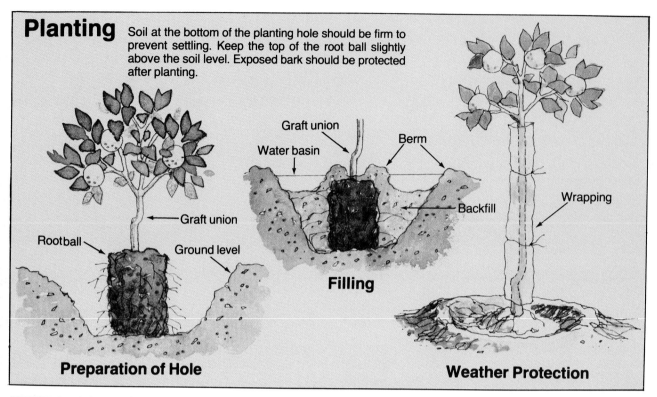

Planting Soil at the bottom of the planting hole should be firm to prevent settling. Keep the top of the root ball slightly above the soil level. Exposed bark should be protected after planting.

Preparation of Hole — Graft union, Rootball, Ground level

Filling — Graft union, Water basin, Berm, Backfill

Weather Protection — Wrapping

WHERE TO PLANT

Citrus trees can be planted indoors and outdoors, in containers that suit the variety, and directly in the soil. Whatever the case, the best site is one that is wind-free and has a full southern exposure that provides maximum sun and heat. Under desert conditions, however, heat is not at a premium, so choose the site accordingly. If you intend to plant directly in the soil and have any doubts about the suitability of a site, leave the tree in the nursery container at the site for a time to see if it is adequate for the tree.

PLANTING STEP BY STEP

Citrus can be planted any time of year, but spring is best. During winter the danger of frost exists and the tree may not be properly acclimated. Planting in spring allows a full season for the tree to become established in the new location before cold weather.

Dig the hole to the depth of the rootball and about twice the diameter. Save the soil from the hole for backfill.

Carefully remove the tree from its container and place it in the hole. If it is a metal container, you may want to have it cut at the nursery. If you do have the can cut, plant the same day. If there are any circling or matted roots, spread them out or cut them off. Be careful not to drop or break the rootball.

It is most important to plant at the same height or slightly above the height of the soil level in the nursery container. Moist soil against the trunk above the origi-

nal soil line can help cause gummosis disease or other disease problems.

Most citrus is sold in containers. Citrus trees headed for commercial orchards are usually available in "sleeves," long, narrow containers approximately six inches by two feet. If yours is a balled tree wrapped in burlap, fill the hole around the tree two-thirds full, untie the top twine, fold back the burlap and finish filling the hole. The burlap will gradually decay.

Citrus trees purchased through a mail-order supplier may arrive bare root. To plant these, make a conical mound in the bottom of the planting hole and spread the roots over it. Fill gradually, compressing the soil with your hands every few inches.

The best backfill soil is the soil the established tree will have to grow in. If the soil has been heavily amended with huge amounts of organic matter, the newly planted tree may settle. Firm the soil in the bottom of the hole to prevent settling.

After the tree is planted, make a soil basin just a little larger than the rootball. Water thoroughly and check the original soil line one last time. If the tree does settle, now is the time to move it back to the correct position with the soil level against the trunk the same or slightly lower than it was in the container. Enlarge the watering basin as the tree grows.

A good rule is to not put fertilizer of any kind in the planting hole. Wait until the tree has put out at least one root flush before applying concentrated inorganic fertilizers. Otherwise you risk injuring the roots.

Watering

There are many ways to water citrus trees. The method you choose will depend on where and how your trees are growing. It is important to note that citrus feeder roots extend beyond the drip line.

Drip line

Water here

Water here

Furrow

Container

Emitter

Drip hose

Basin

Emitter

Drip hose

Emitter

Fertilizers are more safely applied to the soil surface after the tree has been planted, and then only sparingly.

WATERING

Timely watering is essential for proper growth and fruiting of citrus. Without the appropriate water supply you cannot expect satisfactory tree performance. More trees are stunted or lost by drought stress than from any other cause.

To ensure adequate moisture the soil should be thoroughly wet before wilting occurs. To avoid overwatering, excess water must drain away. Hence the need for a soil that drains well.

There is no special way to apply water to a well-drained soil. Any conventional method will work as long as the soil does not remain saturated. Alternate wetting and drying allows oxygen necessary for root growth to enter the soil. Saturated, airless soil creates an unfavorable environment for beneficial soil organisms and encourages disease.

In some desert areas, water is very salty so overhead sprinkling should be avoided.

Soils which are dry when freezes occur increase frost hazard. Wilted drought-stressed citrus suffer more from frost than those that are not under stress.

Very hot days, hot winds or any weather conditions that extract water from the leaves faster than the roots can draw it from the soil, cause temporary wilting, even if ample soil moisture is present. Such conditions promote blossom and small fruit drop, and larger fruit may be reduced in size or drop. Leaves may be injured or lost.

Watering correctly is a skill. No one should be more expert on watering your trees than you. There are too many variables for anyone to give an exact, calendar-regulated set of rules. But we do have these suggestions:

• Young trees need special attention, especially just after planting. As a young tree is establishing its roots in the surrounding soil, which often is somewhat different in character, the rootball can actually dry out while surrounding soil is wet. In warm weather, a young tree may need water as often or more often than once a week.

• Citrus in containers generally need more frequent watering than those in the ground. With containers, overwatering is much less likely. If you have used good soil there will be sufficient drainage and aeration.

• Hot and windy weather means more frequent watering. Conversely, cool overcast days place little demand on the soil's water supply.

• Keep the area around the base of trees free of weeds.

• In areas of salty soil or water, irrigate the root area thoroughly each time you water. A stick or soil probe can help determine how deep the water is going. Moisture to a depth of four feet or beyond is ideal. Let a hose run very slowly for hours.

Citrus trees are susceptible to many trunk and bark diseases. If possible, keep a lawn-free basin around the trees at least the size that is around these grapefruits.

MULCH

A mulch is an insulating layer of plant residue or other material that covers the soil. Mulches are beneficial. They conserve moisture, control soil temperatures, prevent surface compaction or crusting, reduce water runoff and erosion, improve soil structure and control weed growth.

Common mulching materials include compost, sawdust, wood chips, shredded prunings and straw. Paper, plastic, sand and gravel are occasionally used as a mulch. Some mulches such as stone reflect heat and light which is beneficial to the tree.

Keep any mulch that tends to stay wet a few inches away from the trunk. Citrus is very susceptible to bark diseases encouraged by moisture at or near the soil level.

CITRUS IN THE LAWN

A lawn is not the best place for a citrus tree. Grass generally needs more water and fertilizer than is good for the tree. With water and fertilizer kept to a minimum, lawn and tree will compete for the little available, to the detriment of both.

Never try to maintain the lawn right up to the tree's trunk. Damage from lawn mowers, inevitable in such a situation, is very harmful.

If you already have a citrus tree in the lawn, it is best to maintain a wide buffer zone of open, mulched soil from the trunk to somewhat beyond the drip line.

Fertilizing

There are at least 12 mineral elements citrus trees need in order to be healthy. The familiar nitrogen, phosphorus, potash, magnesium, calcium and sulfur are necessary in relatively large amounts. Micronutrients, needed in minute quantities, are iron, zinc, manganese, molybdenum, copper and boron.

Most California and Arizona soils contain adequate supplies of all these nutrients except nitrogen. Thus, one could fertilize with only a nitrogen fertilizer and expect good results. The other nutrients are necessary if deficiencies occur.

Nutrient deficiencies and excesses can often be detected by closely observing the leaf size, color, shape, and any patterns that may develop. Carelessly used fertilizers or herbicides may produce leaf symptoms similar to mineral deficiencies or excesses. Also, limb "sports," a naturally occurring genetic mutation, show distinctive color patterns but are rather easy to distinguish from normal leaves and fruit.

THE ROLE NUTRIENTS PLAY

Nitrogen—Nitrogen stimulates growth and is essential to fruit setting. Bearing citrus trees need nitrogen most during flowering and fruit set. Excess nitrogen late in the season causes coarser fruit, sensitizes the tree and fruit to cold and delays fruit maturity. Contrary to some popular notions, nitrogen cannot make fruit grow larger. Only proper irrigation will do that.

Phosphorous—Most soils contain ample phosphorous to meet the citrus tree's needs. Because the absorption rate is slow, phosphorous toxicity is unlikely, although an excess in the soil may make other nutrients unavailable to the tree. Phosphorous deficient trees produce coarse fruit with a thick rind. Severe deficiency results in misshapen fruit and reduced yields. Such trees are less vigorous and have less spring bloom. Treble super phosphate is a good fertilizer for deficient trees.

Potassium—Citrus roots absorb potassium readily. A good balance of potassium helps enhance fruit quality. A significant deficiency results in smaller fruit with a thinner rind, making it more prone to fruit splitting and premature drop. A significant excess produces a coarser fruit of higher acid content and delays the maturity of the fruit. Most soils are amply supplied with potassium, though it may be flushed from container soils.

Magnesium—Magnesium plays a major part in the development of fruit color, both internal and external. Poor growth characteristics as well as increased cold susceptibility result from a lack of magnesium. Defi-

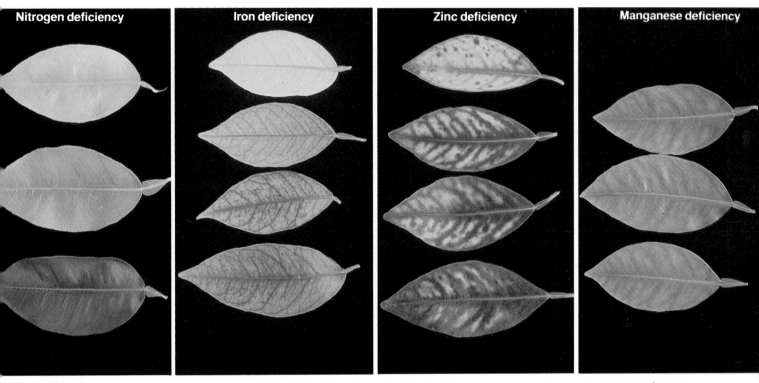

Nitrogen deficiency | **Iron deficiency** | **Zinc deficiency** | **Manganese deficiency**

Nitrogen deficiency causes yellowing in older, lower leaves first. The plant's ability to absorb iron, zinc and manganese is more important than quantity present in the soil. Soil pH, water conditions and temperature strongly affect their availability. Deficiency symptoms of iron, zinc and manganese generally occur in newest growth first.

ciency symptoms commonly develop in the late summer or fall, particularly with container grown citrus. Deficiency is aggravated by excessive irrigation or manure applications. Magnesium hunger may be relieved by magnesium sulfate or dolomite lime, but leaf sprays of magnesium nitrate are effective and economical. Magnesium chelate is another recommended magnesium source.

Calcium–Calcium is essential to all plants. It is rarely deficient in the West because soils there are naturally high in calcium. Deficiency symptoms include death or stunting of growing tips, premature shedding of flowers and weakened stems. It may be a problem in container soils. A good fertilizer source of calcium is calcium nitrate found in gypsum. If soil is acid, use ground limestone or dolomite.

Sulfur –Sulfur may be deficient in some California soils, notably the eastern foothills of the San Joaquin Valley and the central coastal range and the Sacramento Valley. Symptoms of sulfur deficiency include a light, yellowish green leaf color that is not improved by nitrogen fertilizer. Growth will be generally less vigorous. Many nitrogen fertilizers such as ammonium sulfate contain enough sulfur to correct any deficiency that may occur. Gypsum is a good source of both calcium and sulfur.

Manganese–A deficiency will cause reduced growth and less fruit. Apply sprays containing manganese to quickly correct this.

Zinc–Sooner or later most citrus trees require an added dose of zinc. Zinc deficiency reduces growth and fruitfulness. Soil applications of zinc are not usually effective once the tree shows the symptoms—interveinal chlorosis, twig dieback, rosetting of terminal leaves. Foliage correction sprays are best. Zinc sulfate is the recommended corrective.

Copper–Copper deficiency is rare. Lack of copper reduces fruit production and fruit quality. Copper-based fungicide sprays normally contain enough copper to correct deficiency problems that may occur. Oranges may form gum pockets around central pith if copper is deficient. Copper deficiency sometimes occurs in container-grown citrus growing in soils with a high proportion of peat.

Iron–Alkali and excess salinity contribute to iron deficiency; over-irrigation aggravates it. Iron deficient trees grow poorly and production suffers. Proper irrigation is often the best treatment. Chelated iron compounds applied to the soil help but vary in effectiveness due in part to the fact that different citrus rootstocks vary considerably in their ability to absorb iron. A chelated iron spray is usually the best solution.

When to Fertilize

 Spread fertilizer on soil.   Spray foliage.

January	February	March	April
In soil: 🖐	🖐	🖐	🖐 Micronutrients if needed
In container: 🖐	🖐	🖐	Micronutrients if needed 🖐
May	**June**	**July**	**August**
In soil: 🖐			
In container:	🖐	🖐	
September	**October**	**November**	**December**
In soil:			
In container:			

Boron — Excesses have occurred in orchards where the irrigation water is naturally high in boron. When present in toxic amounts, boron causes an irregular browning of the leaf margins. If deficient, leaves are thickened, curled, wilted and chlorotic. Branches tend to grow into a "witches' broom," or a mass of small twigs.
Molybdenum — Required by all plants, it is rarely deficient in the West. Deficiency in molybdenum, however, sometimes causes a condition known as yellow spotting.

HOW MUCH FERTILIZER?

Recommended rates of nitrogen are expressed in pounds of "actual" nitrogen. Mature citrus trees planted in open soil usually require 1 to 1½ pounds of actual nitrogen per year.

Most fertilizers are not labeled according to pounds of actual nitrogen, but the *percentage* of actual nitrogen of the bag's contents. For example, a good citrus fertilizer labeled as 12-6-10 contains 12 percent nitrogen, 6 percent phosphorous and 10 percent potash.

To convert the percentage nitrogen to pounds of actual nitrogen, multiply the percent nitrogen by the total pounds of fertilizer in the bag. A 10 pound bag of 12-6-10 would contain one and two-tenths pounds of actual nitrogen.

The following guide can help determine how much to fertilize young citrus trees growing in the ground.

Age of tree in years	Pounds of actual nitrogen per year:
1	.1
2	.2
3	.35
4	.5
5	1 to 1.5

WHAT KIND OF FERTILIZER?

Any fertilizer that supplies adequate nitrogen can be used for citrus. Nurseries carry products specifically labeled "Citrus Food" or "Citrus and Avocado Food". These are called complete or balanced fertilizers and are formulated for citrus and presumably contain all necessary nutrients. The application rate on these fertilizers is usually determined on the same number of pounds of actual nitrogen recommended above.

Commercial-grade fertilizers, such as urea and ammonium nitrate, supply only nitrogen. Ammonium sulfate contains nitrogen and sulfate. These are more economical than a complete fertilizer if larger quantities are needed.

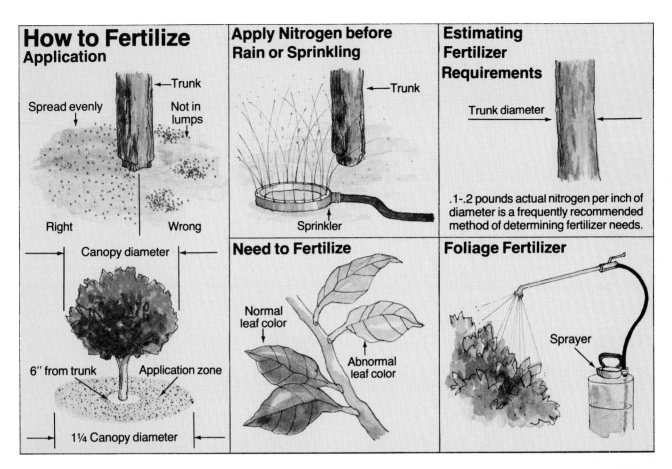

How to Fertilize
Application

Spread evenly — Trunk — Not in lumps

Right — Wrong

Canopy diameter

6" from trunk — Application zone

1¼ Canopy diameter

Apply Nitrogen before Rain or Sprinkling

Trunk

Sprinkler

Estimating Fertilizer Requirements

Trunk diameter

.1-.2 pounds actual nitrogen per inch of diameter is a frequently recommended method of determining fertilizer needs.

Need to Fertilize

Normal leaf color

Abnormal leaf color

Foliage Fertilizer

Sprayer

WHEN TO FERTILIZE

Depending on the soil type and growing conditions, divide the total year's application into three or four lots. More frequent applications are needed in sandy soils or heavily watered soils. Apply the first lot in January or February and the balance at four to six week intervals. We advise you not to fertilize later than late August or early September. Excess or late applications of nitrogen encourages new frost-sensitive growth and adversely affects fruit quality.

HOW TO FERTILIZE

Spread fertilizers evenly over the entire root area. Do not place it in lumps or piles which can burn roots. Apply just ahead of a good rain or prior to irrigation. Then it can wash into the soil and go to work at once. Nitrogen loss is considerable if the fertilizer remains on the soil surface, particularly in hot weather.

FERTILIZING CONTAINER CITRUS

Citrus grown in containers need particularly close attention to nutrient supplies. Unlike natural garden soil, container soils or potting mixes need more frequent replenishing of necessary elements. More frequent waterings necessary for container citrus also flush nutrients, especially nitrogen, out of the soil.

To compensate, use a complete fertilizer and follow the manufacturer's recommendations regarding container citrus. Generally such instructions call for more frequent applications of smaller quantities.

Micronutrient deficiencies are most likely to occur with container citrus. If chlorosis (yellowing) and mottling are not corrected by the usual applications of nitrogen, the deficiency is probably manganese, zinc or iron. Many balanced fertilizers contain chelated forms of these nutrients which should correct the problem.

Wet winter conditions often cause chlorosis in spite of a good fertilizer program. It is a condition frequently seen in container-grown citrus. In a season of high rainfall, the soil remains wet for long periods and essential elements are leached out or roots rot away and become unable to absorb needed mineral elements. During the winter and early spring months, warm spells may occur while the roots are still inactive in a wet or boggy soil. Top growth occurs without the necessary elements, causing the chlorotic appearance. Further, roots in containers are not as insulated from cold temperatures as roots in open soil and low temperatures alone can inhibit root activity. If chlorosis following a cold period is common, a fall application of micronutrients might be helpful.

APPLYING MICRONUTRIENTS

The most common micronutrient deficiencies in citrus are iron, zinc or manganese. These elements may be supplied by soil applications but are best corrected by spraying the foliage. Response is much more rapid. These nutrients are available in sulfate forms, such as zinc sulfate. Apply as directed on the package label. Chelated forms of these elements are available as nutrient sprays and have proven safer and more efficient. Chelated forms of micronutrients can be used either as sprays or broadcast around the base of the tree and watered in.

The best time to apply micronutrient sprays is in spring when the leaves of the first flush of growth are two-thirds expanded. Apply in fall to plants in containers to increase the plant's reserves and help prevent winter and spring chlorosis. Older leaves do not take up the nutrient spray as readily. If leaf symptoms indicate acute deficiencies, spray at once, perhaps repeating at a more favorable time.

TRUST YOUR EYES

Close observation of leaf appearance is the most convenient and practical way to manage your nitrogen fertilizer program. Any deviation from normal color during the growing season calls for changes in water or fertilizer applications. If the tree does not respond, it could be the result of a micronutrient deficiency.

Pruning

Citrus trees need little pruning. When a tree is mature, pruning should be confined almost entirely to removing dead, diseased or broken branches. Pruning is more important for commercial growers for simplifying harvesting, spraying and other operations. When the trees crowd excessively, some pruning is necessary to admit light to convenient fruiting areas. Remember, because citrus stores its food in its leaves, the amount of foliage is directly related to the amount of fruit the tree will produce.

In the home garden, appearance of the tree is as important as fruit production. Young trees will occasionally put out a vigorous branch that gives the tree an unshapely appearance, but ultimately the tree will fill out to its full characteristic shape. Rather than wait until this occurs naturally, you may chose to lightly prune overly vigorous shoots and remove trunk suckers while they are small. Any suckers originating from below the bud union should be removed. As the

tree matures, the need for this treatment will diminish. Unless you wish to train your tree to a high branching structure, do not remove low-hanging "skirt" branches. They bear fruit which are easy to reach and they shade the ground, which prevents weed growth. A full skirt of foliage also helps reduce sunburn on tender bark and fruit. This is especially important in hot inland and desert areas.

Lemons need more pruning than oranges or grapefruit. The latter two are twiggy and grow eventually into a pleasing shape. Lemons tend to grow in a more vigorous, upright manner. It may be desirable to remove or cut back some vigorous shoots of young lemon trees. A moderate annual thinning and heading back will reduce the crop somewhat but result in a more attractive, bushier tree. After a time the lower, fuller tree will produce more fruit that is easier to reach.

Citrus can be pruned to suit your landscape needs, but fruit harvest will be somewhat diminished.

REJUVENATION PRUNING

In many parts of the western United States, old citrus groves are subdivided into home sites. Often trees remain that the owner would like to restore. Beautiful landscapes have been created by restoring individual trees or entire orchard blocks. Proper pruning can do much to invigorate old or weak citrus trees. Of course, the severity of the pruning depends on the condition of the tree.

First, remove damaged, diseased or dead wood.

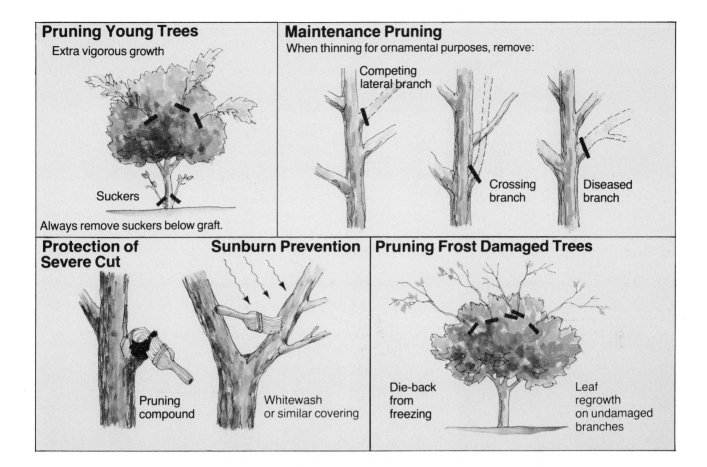

Pruning Young Trees
Extra vigorous growth

Suckers

Always remove suckers below graft.

Maintenance Pruning
When thinning for ornamental purposes, remove:

Competing
lateral branch

Crossing
branch

Diseased
branch

**Protection of
Severe Cut**

Pruning
compound

Sunburn Prevention

Whitewash
or similar covering

Pruning Frost Damaged Trees

Die-back
from
freezing

Leaf
regrowth
on undamaged
branches

This will probably expose some healthy branches and trunk directly to the sun, inviting sunburn. Cover such newly exposed wood with a coat of indoor latex white paint. Large pruning cuts may be treated with an asphalt emulsion pruning compound.

In the most severe cases rejuvenation requires complete removal of all the branches except for the main scaffolds, the basic framework branches, leaving almost no foliage. All exposed branches and major cuts should be protected as mentioned above.

After any heavy pruning apply white latex paint to newly exposed bark and begin a water and fertilizer program that will restore the tree's vigor.

PRUNING FROST DAMAGED TREES

The most important thing to do when faced with a frost damaged tree is, wait. Do not do anything until you know the extent of the damage. Wait until new growth clearly defines the damaged areas. This may mean waiting as long as six months after the freeze occurred.

Symptoms of minor frost damage are yellow, droopy or wilted leaves. Frozen leaves may not change color but shrivel and drop before they dry. In some cases, they dry and shrivel on the twig and may remain on the tree for several weeks. When they finally fall, naked dead twigs persist on the exposed portion of the tree. The last growth cycle of the fall season often defoliates and dies, leaving blackened young shoots.

If the extent of the injury is only to twigs and small limbs, you can prune as soon as the danger of frost is past. Always allow time for new growth to begin. If you prune too early some limbs will continue to dieback and you may have to redo the job.

Occasionally winter frosts are severe enough that citrus trees suffer serious injury. It may be impossible to determine the extent of the injury for several months. In the worst cases, dieback may continue through the entire growing season following the freeze. During this period little can be done and treatment should be postponed. In the meantime, remove fruit of no value, limit fertilizer to a minimum and be very careful not to overwater.

PROTECT YOURSELF

Many varieties of citrus have formidable thorns. Protect yourself with heavy gloves and goggles. Use a strong ladder if one is needed. Use proper cutting tools that are in good condition to avoid struggles that lead to accidents.

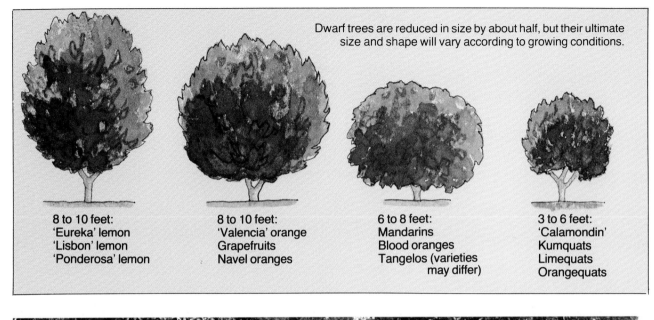

Dwarf trees are reduced in size by about half, but their ultimate size and shape will vary according to growing conditions.

8 to 10 feet:
'Eureka' lemon
'Lisbon' lemon
'Ponderosa' lemon

8 to 10 feet:
'Valencia' orange
Grapefruits
Navel oranges

6 to 8 feet:
Mandarins
Blood oranges
Tangelos (varieties
    may differ)

3 to 6 feet:
'Calamondin'
Kumquats
Limequats
Orangequats

A dwarf version of a 'Minneola' tangelo: It is five years old and within three to five feet of its approximate full size.

Growing Dwarf Citrus

Dwarf citrus planted in the ground will normally achieve a height and width of about eight feet and will be globular in shape. Although this description fits many types, individuals may develop variable forms.

Dwarf citrus are usually 40 to 50 percent of the size of standard orchard trees. Because standard citrus trees vary in size depending upon the variety, dwarf citrus varies proportionately. A 'Valencia' orange or 'Marsh' grapefruit become 18- to 20-foot trees on standard roots; a kumquat grows to only 8 to 12 feet. Dwarfs of these varieties will be about 8 to 12 feet and 3 to 6 feet, respectively.

Every aspect of the tree's environment—air, temperature, humidity, soil, water quality, planting density and nutrition—influences the tree's performance and ultimate size. Experimental dwarf citrus have, with ideal conditions, grown larger than those shown in the illustration. The planting density and resulting root competition, typical in the home garden, often prevent trees from reaching their maximum size.

A standard-size version of a 'Minneola' tangelo: At 15 to 20 years old, it has achieved maximum size.

Dwarf citrus produce normal fruit. Interestingly, dwarf citrus have proven to make proportionately more fruit than their standard size counterparts. Dwarfs produce 50 to 60 percent of the fruit of a standard size tree, but in only one-quarter the space. This is an advantage to the home gardener. He can have four varieties in the space of one full-size tree.

Some varieties of citrus such as 'Meyer' and 'Ponderosa' lemons are natural dwarfs. They are often propagated and sold as rooted cuttings. Citrus buyers should be aware that other than these naturally dwarfed types, cutting-grown citrus are not necessarily dwarf citrus. A dwarf is usually budded or grafted onto a specific rootstock.

Citrus propagated as cuttings are often sold as shrub or container citrus, but may ultimately be almost as tall as standard citrus.

The common rootstocks used by dwarf citrus growers include the trifoliate orange, 'Cunningham' citrange and the 'Cuban' shaddock. Because there is no universal rootstock, the optimum rootstock-scion combination depends upon the growth characteristics of the variety to be propagated and the ease with which they are grown.

Citrus Spoilers— Pests, Diseases and Problems

Homeowners usually experience relatively few citrus pests that cause serious problems. That is why a preventative spray program is unnecessary, unlike most deciduous fruit trees. By growing healthy trees with proper water and fertilization, pest damage will be further minimized.

Occasional problems may occur which warrant some control measures. But consider how much damage is "tolerable" before using an insecticide. Many of the most potentially damaging citrus pests are kept in control naturally by various predators and parasites. Frequent use of an insecticide will likely be more upsetting to those beneficial insects than to the pests.

The home gardener's citrus pest control program should first rely on these same natural enemies. If you have only one or a few trees, you, the gardener, may be the most efficient natural enemy. You can wash the trees clean or hand-pick a few worms.

Citrus trees are very appreciative of frequent "showers." Hosing them down daily or as frequently as you can manage keeps humidity high and helps discourage many pests. Adding a small amount of dish soap to the water (two teaspoons to a gallon) will help stop many pests.

Citrus insect pests most likely troublesome to the home gardener are illustrated on the following pages.

CITRUS DISEASES

Citrus is an important commercial crop, so more than 80 diseases have been identified and intensely studied. Many are significant on the large scale but rare or unimportant otherwise. Following are descriptions of those citrus diseases that are most frequently problems in home gardens.

Gummosis or foot rot. This disease begins at or near the ground level. Dead and decayed patches of bark indicate its presence. Large amounts of gummy sap usually flow from the infected areas. It usually occurs in winter where wet soil or standing water remains in contact with the bark for days or weeks at a time. Trees in clay soils may be more prone to the disease because such soils drain more slowly. Trees properly planted and cared for are usually not susceptible.

A copper-based fungicide can prevent the disease if sprayed on the trunk. If the disease is a common problem in your area and winters are particularly wet,

Gummosis (foot rot) is evident by dead bark and gummy sap in infected area.

a preventative spray may be advisable.

If a tree shows symptoms, remove the soil down as far as the bark is diseased to determine the extent of the damage. If the bark has been killed more than half way around the trunk, it will be best to remove the tree and plant another one. If half or more of the bark is still sound, carefully use a knife to remove the part which has been invaded by the fungus and one-quarter inch of unaffected bark around the margins. Bark which is alive may be yellow and gummy next to the wood. Remove only parts that are brown and discolored. Do not scrape the exposed healthy tissue of the wood; removal of the diseased bark is the only cutting necessary. Disinfect the wound with a solution of potassium permanganate, two teaspoons per quart of water. When the bark begins to form a healing callus over the wound, cover the wound with an asphalt emulsion, white lead paint or similar non-toxic material.

If the same fungus splashes from the soil onto fruit, it will cause the disease *brown rot*. Fruit losses from brown rot are often more economically significant than losses due to gummosis. It may be prevented by a zinc-copper fungicide sprayed around the lower leaves just before the rainy season.

Oak root fungus. The *Armillaria mellea* or oak root fungus, is often present where oak and other native trees have been removed or where flood waters have deposited infected wood. It causes a moist decay of bark and wood and can be identified by the white, fan-shaped fungus growth which develops in and under the bark, or by dark, root-like strands growing on the surface of roots. During late fall or early winter, groups of light brown toadstools may be produced.

All citrus rootstocks are susceptible, and trifoliate

and Troyer citrange seem particularly sensitive. The disease may be retarded by exposing the crown roots and trunk so they remain dry. No cure is known, so removal of seriously infected trees is recommended.

Dry root rot affects all rootstocks when conditions are favorable. Sudden wilting occurs when most of the root system is involved. The roots show moist decay that later dries, cracks and shreds. A dusky to black discoloration of the wood may be revealed by cutting through the roots.

Damage to roots by excess fertilizer, pesticides or rodent activity provides entrance for agents causing dry root rot. To prevent trouble from this disease, avoid all types of root injury.

Virus disease. If you live in an area of commercial citrus orchards or have a grove of your own, you should be aware of the potential for virus disease. Typical symptoms are poor growth or an unusual mottling of leaves, bark or fruit. Whenever problems persist regardless of your efforts to control them, suspect virus disease. Check with a local citrus authority if you suspect this problem.

Tristeza or quick decline. Tristeza is a disease that has killed huge numbers of trees in many parts of the world. Most susceptible are trees on sour orange rootstocks. Rootstocks with some resistance are rough lemon, 'Troyer' and 'Carrizo' citranges, trifoliate orange and 'Rangpur' lime.

Above-ground symptoms of affected trees are similar those produced by oak root fungus or gopher-riddled roots. Leaves become a dull ashen color and curl lengthwise and upwards. This above-ground deterioration parallels root destruction.

There is no real remedy for tristeza. The disease is the reason for the many state and county citrus quarantines. 'Meyer' lemons were typically infected and spread the disease via an aphid transmitter. All the old 'Meyer' propagation material was destroyed and replaced with virus-free 'Improved Meyer' lemon. If you suspect a tree of yours has been infected with this disease, consult a County Extension Agent.

Citrus tree dying from Tristeza (quick decline). Healthy trees in background.

Citrus blast. This disease is unique to citrus in the Sacramento Valley of California. Winter-damaged leaves or twigs are infected by a bacteria. Infection typically starts in the crease between the leaf blade and leaf stem. As the disease progresses, the twig is girdled and dies. The result is a tree with many persistent dead leaves or "flags." Usually the worst damage is on the south and east sides and tree top. Copper sulfate mixed with hydrated lime is the usual control when one is necessary. Windbreaks, winter-ready trees and removal of previously killed twigs is usually sufficient to prevent this disease.

Gumming. This problem is caused by some kind of mechanical injury such as a lawnmover bumping the trunk. Sap oozes from a crack or fissure in the bark and makes a gummy deposit on the limb or trunk. It is not a disease so fungicide treatment is not necessary. It may be helpful to make a few holes in the bark to allow the sap to drain.

Sunburn. Citrus bark is thin and easily damaged by the sun. Young trees without a dense foliage cover are most susceptible. Sunburned bark is hard and brittle and may peel off in patches. When the trunk is exposed, it should be protected in some manner. The common way is to use a whitewash sold for that purpose. Protective wraps are best for young trees. Damaged areas should be whitewashed to prevent further injury.

Fruit drop. Normally a fairly heavy drop of small fruit begins shortly after blossoms fall. This continues until the fruit has grown to about one-half inch in diameter. The amount of spring drop depends on weather and tree conditions. Excessive drop may be caused by lack of moisture, too much available nitrogen fertilizer, sudden heat spells and insect or spray damage. Cooling the trees in hot spring weather reduces drop significantly. Appropriate watering, fertilization and pest control will minimize fruit drop.

Leaf drop. Citrus trees are evergreen but they shed some leaves throughout the year. Lack of water, nitrogen, insects and frost injury are some other causes of leaf drop.

Fruit splitting. Changes in weather are the usual cause of fruit split. High humidity after a dry period is usually enough of a trigger. However, only a few fruit on the tree are normally affected.

Gophers. Every citrus orchard we know maintains an ever-vigilant gopher control program. These varmints can kill a tree in a single night. Poison and trapping are the common controls. Dig down to a main tunnel and set spring traps in both directions. Gopher control services are available in most citrus growing areas.

Citrus Pests

APHIDS

Aphids typically attack lush, new growth and cause the leaf curling and stunting shown in the photograph. More seriously, aphids may transmit a virus disease known as *tristeza,* or quick decline (see photo page 133). Ants feed on the syrupy-sweet aphid secretion called *honeydew.* They move the aphids throughout the tree and often protect them from predators.

Remedy. It is usually best to rely on natural predators and parasites such as lady beetles and lacewing flies. Young trees can be washed clean of aphids with a spray of water or soapy water.

Aphids appearing on underside of leaf cause leaf curling and stunting.

MEALYBUGS

Mealybugs are soft-bodied insects closely related to both aphids and scale. They can move around, but generally settle in clusters on the leaves and twigs and at the angles where fruits touch. Like aphids, they secrete a sticky honeydew that is encouraged by ants and may promote growth of black sooty mold.

Remedy. Mealybugs are primarily controlled by natural enemies, most important of which is the mealybug destroyer. It is a ladybug type of beetle that is mostly shiny-black but has a red head and red wing tips. Washing and soap sprays are sometimes used for indoor citrus. Oil sprays and Malathion are recommended for particularly heavy infestations.

Mealybug, left, secretes sticky honeydew.
At right, clusters of mealybugs on twigs.

SCALES

The scale insects are usually considered the most troublesome citrus pest. California red scale, Florida red scale and Yellow scale are very similar and usually most damaging. Others are brown scale, dictyospermumscale, black scale, purple scale and cottony cushion scale. California red scale, considered the most important citrus pest of California, infests twigs, leaves and fruits. Florida red scale and yellow scale are important citrus pests in the Gulf States and Florida and infest only leaves and fruits.

Remedy. There are natural enemies of these scales, but their capacity to control the pests is variable. Cottony cushion scale is usually well controlled by the Australian lady beetle. Brown scale and black scale are also usually biologically controlled. Red scale is controlled by parasites in some areas, but not others, even though the parasite is present.

Watch to determine if parasites (naturally occurring or introduced) are helping curtail a scale outbreak. If they are not, you have two alternatives: One is an "oil" spray. Oil sprays are specially prepared petroleum oils that coat the insect long enough to smother it. Oils appropriate for citrus are classified as "light-medium." Use in fall when temperatures are moderate. Do not use oils if temperatures are high, if trees need water, or if freezing temperatures are expected within 24 hours.

The other alternative is an insecticide such as Malathion. Spray after bloom and before scale become established on fruit surfaces. The label will include other specific instructions. Malathion and light-medium oils are sometimes combined to control citrus scale pests.

Top left, California red scale. Top right, purple scale.
Bottom left, soft scale. Bottom right, black scale.

CITRUS RED MITES

Of the variety of spider mites that infest citrus trees, the citrus red mite is most damaging. They can usually be found year round but populations are lowest in winter and midsummer. The mites scrape tissue from leaf undersides and remove chlorophyll, creating a silvery, speckled look on the leaf top. Severely damaged leaves become silver, then turn brown before drooping. The Texas citrus mite causes leaf damage almost identical to the citrus red mite. The citrus bud mite is a pest of lemons in California. It becomes established in the buds and blossoms, causing odd deformities of leaves and fruit.

Remedy. Frequent washing and high humidity discourages spider mites. Once populations are high, use a light-medium citrus oil or a recommended miticide such as Kelthane.

Left: Citrus red mite on leaf; Right: speckled leaf damage.

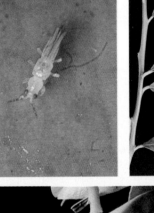

EUROPEAN BROWN SNAILS

A familiar pest in most gardens, snails are particularly fond of citrus. In citrus areas of coastal California snail populations can reach alarming levels (see photo). They make irregular holes in citrus leaves, pits or scars on fruit and can completely cover the trunk with their shells.

Remedy. Poison baits are the most common recourse. Those containing Metaldehyde are safer; Mesurol® is somewhat more effective, but requires more caution. Liquid sprays of both of these materials are also available. The decollata snail is predatory on the European Brown Snail to a significant degree. Frequent inspection and hand-picking can be very effective in small gardens.

Left: Huge snail population covers tree trunk.

Right: Snail damage to leaves.

CITRUS THRIPS

Thrips are among the most important citrus pests of the Southwest. Citrus thrips survive winter in the egg stage on the stems and leaves of infested citrus trees and sometimes pepper trees. Young thrips, or nymphs, begin to appear around March.

Damage includes distortion of new growth and leaves, and fruit scarring. A sure sign is a definite ring around the fruit at the blossom end.

Remedy. Thrips are one of the toughest pests to control. Timing is important. If you have a thrip problem every year, spray immediately when flowers drop. Don't wait until you can find them or their damage. Diazinon spray should help but sprays recommended for thrips change frequently. Consult with a local citrus authority for the latest information.

Top left: Citrus thrip (magnified approximately 25 times) on leaf.

Top right: Thrip damage causes distortion to leaves.

Bottom: Fruit damage caused by citrus thrips can be seen as tight rings circling the fruit near blossom end.

Texas is grapefruit country.

The 'Star Ruby' grapefruit, a Texas speciality, has more color than any other and is often made available by mail order.

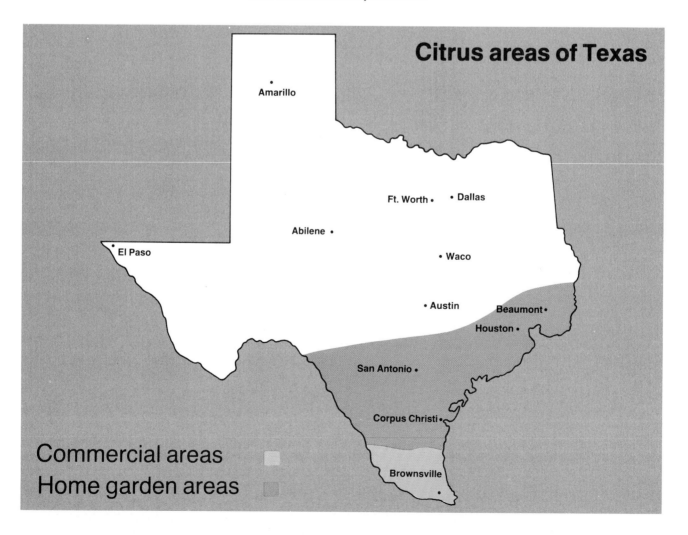

Citrus areas of Texas

- Amarillo
- Ft. Worth • Dallas
- Abilene •
- El Paso
- • Waco
- • Austin
- Beaumont •
- Houston •
- San Antonio •
- Corpus Christi •
- Brownsville •

Commercial areas
Home garden areas

Growing Citrus in Texas

Most details of citrus culture are universal and useful to citrus gardeners everywhere. But climate always has a considerable influence. Here we switch consideration from the primary citrus areas of California and Arizona to Texas.

THE HISTORY OF CITRUS IN TEXAS

The history of Texas citrus industry is a study of "man against the elements." It is a story of repeated setbacks caused by killing frosts. Each time, however, the industry has managed to come back with great determination.

Unlike California, which is somewhat protected from arctic air by mountain ranges, Texas lies in the infrequent but direct path of frigid air masses extending down from the north. As one Texan put it, "There is nothing between us and the North Pole except barbed wire fences."

Citrus was brought into Texas from the Gulf Coast. The first trees were planted in the Houston-Beaumont area in the 1890's. Satsuma mandarins were the main variety, but most of these trees were destroyed in the freezes of 1894-1895 and 1899. Despite this, planting continued. By 1910 the United States census listed 833,406 citrus trees planted in Texas. These were primarily Satsuma mandarins in the Beaumont and Houston areas. The 1916-1917 freeze destroyed the majority of these trees and the 1920 census showed only 123,951 citrus trees remaining in Texas.

The commercial industry gradually moved south into the lower Rio Grande Valley. Primarily oranges and grapefruit were planted and within ten years the industry was again thriving. In 1949 and 1951 severe freezes struck again and destroyed 80 percent of the citrus trees in the state.

With amazing determination and resiliency, growers began to replant almost immediately. Just as the work was beginning to pay off, another freeze in 1960 killed 30 percent of the trees.

Today, the citrus industry is once again thriving, and the primary commercial production still resides in the lower Rio Grande Valley. More modern techniques of frost protection are being employed, but the threat of severe damage is a very real one.

Early varieties such as 'Marrs' and 'Valencia' are decreasing in planted acreage due to competition from Florida and California. Texas is grapefruit country—1979 crop estimates showed 47,800 acres planted in grapefruit and 28,100 acres of oranges. Of the grapefruit, 83 percent were 'Redblush'; 9 percent were the newer variety, 'Star Ruby'. Because of the taste appeal of 'Star Ruby' an increase is expected.

Due to its central geographic location, most Texas citrus is marketed fresh. Some of the premium Texas grapefruit are sold through mail order.

THE TEXAS CLIMATE

Texas, of course, is a very large state and has many different climates. The western part of the state is dry, resembling other arid parts of the West, although the winters are colder. As you move closer to the Gulf Coast and a more southern latitude, the climate becomes more humid, similar to other Gulf Coast states. It is in these humid areas that most of the citrus is grown.

Rainfall is at its highest from May to September, primarily in the form of thundershowers. Because rainfall of this type is so unpredictable, a good watering program is a necessity. Besides the threat of the cold artic air already mentioned, hail from thunderstorms and high winds from periodic tropical storms are occasionally damaging to citrus in the Gulf Coast area.

The warm humid weather in Texas causes citrus trees to grow very fast and citrus fruit to ripen early. For example, a 20-year-old 'Marsh Seedless' grapefruit grown in Texas is usually the size of a 40-year-old tree grown in California. For other effects of a humid climate on citrus see the chapter "Citrus Climates," page 25, which compares the Florida climate, similarly humid, with the drier California climate.

Although the main commercial citrus area of Texas is located in the lower Rio Grande Valley, the area in which the homeowner can successfully grow citrus is much more extensive. It includes the Wintergarden area, northwest of the Rio Grande Valley and a band extending inland about 100 miles (150 km) along the entire length of the Texas Gulf Coast. This land includes the cities of Beaumont, Houston and Corpus Christi.

However, by growing the hardiest varieties, using microclimates and by protecting fruit in winter, the homeowner in colder areas as far north as the Dallas-Ft. Worth area can enjoy citrus and its relatives. Using indoor and outdoor culture techniques (see page 141), anyone in Texas can grown citrus.

WHICH VARIETY?

Texas A&M University makes the following recommendations:

Outside the lower Rio Grande Valley, only cold-hardy varieties should be planted. The most cold-

The 'Nansho daidai' (or *Citrus taiwanica*) is a hardy, natural hybrid of the sour orange of interest to Texas citriculturists. The rind is thick and the flesh is very sour and deep yellow.

The 'Owari' Satsuma mandarin has long been one of the favored citrus varieties in Texas. This south-facing slope provides maximum warmth and cold air drainage.

hardy types of citrus are kumquats, calamondins, Satsuma mandarins, tangerines (mandarins) and trifoliate oranges. If sweet oranges are grown in colder areas outside the valley, only early varieties should be grown so the harvest will be completed prior to any freezing weather. The varieties adapted to the Upper Gulf Coast are marked with an asterisk (*).

SPECIALTY TREES

Many other types of citrus can be used in landscaping, some are edible and others not. Cold tolerance also varies.

Sour orange—Cold hardy rootstock. Produces upright, thorny tree. Fruit rather unattractive but can be used in making candies and acid drinks.

*'Chinotto' sour orange—Mutation of the sour orange used as an ornamental. Has the same degree of cold tolerance. The internodes (distance between leaves) are shorter, making a more compact tree with fewer thorns and more foliage.

'Sinton' citrangequat—Very cold hardy, always red, fruit inedible. 'Thomasville' citrangequat has large, edible fruit.

EARLY ORANGE VARIETIES

'Marrs'—Produces good crops at a young age. Matures September to January.

'Hamlin'—Good producer. Fruit tends to be smaller than 'Marrs' and does not yield as early. Matures October to January.

Navels—High quality fruit but does not produce good crops in most areas. Matures late, September to January.

LATE ORANGE VARIETIES

'Valencia'—Produces fair crops but takes longer to come into production. Matures February to June.

TANGERINES AND TANGELOS

'Clementine' (Algerian)—Cold hardy. Produces good crops when planted with 'Orlando' tanglo as a pollinator. Matures October to January.

*'Fairchild'—Cold hardy. Should be planted with 'Clementine' as a pollinator. Matures November to January.

*'Orlando' Tangelo—Most popular of the tangelos (tangerine-grapefruit hybrids).

*Satsuma—When grown in areas with cold winters,

this tree produces good crops of excellent, flavorful fruit and is cold hardy. Should not be planted south of Corpus Christi.

*'Changsha'—Most cold hardy of the tangerines. Produces good crops of fruit that are rather small and very seedy. This tangerine appears to have considerable potential for landscape use. Seedling plants have produced well with little cold damage in College Station and Arlington.

GRAPEFRUIT

Fruit produced outside the Lower Rio Grande Valley tends to be more acid (sour) than those grown in the valley.

'Ruby Red' or 'Redblush' — Most popular variety in Texas. Grows well and begins to produce at three years of age when cared for properly.

'Marsh White'—Seedless variety. Same growth characteristics as 'Ruby Red'.

'Star Ruby'—New variety is approximately three times as red as the 'Ruby' variety and retains its color later in the year. Trees are more susceptible to foot rot. This variety has a tendency to turn yellow, but usually grows out of it.

ACID FRUIT

'Mexican' lime—Very cold tender. Produces large crops of small (1-inch diameter) excellent-quality fruit. Seedlings and cutting-grown plants will grow back from roots following freezes. 'Mexican' limes are good container plants, often producing fruit for a long period through the year.

*'Meyer' lemon—Produces good crops of large lemons ripening August to March. Plants grown from cuttings are frequently sold for landscape use. These plants will usually sprout from the base after severe cold injury.

*'Eustis' limequat—Lime-kumquat hybrid. More hardy than a lime but less hardy than a kumquat. Makes an excellent small ornamental citrus tree.

*'Calamondin'—Cold hardy. Produces heavy crops of small 1½-inch fruit which can be used as a lime substitute. Calamondins are popular as container plants, indoors and out, and in parts of the country where other citrus are not available or cannot be grown.

*Kumquats—Very cold hardy. Should be grown on 'Cleopatra' mandarin, calamondin or trifoliate orange rootstock. Makes a relatively small, shrublike (10-foot maximum height) tree covered with small, brightly colored fruit. Varieties are 'Nagami', 'Meiwa' and 'Marumi'. 'Meiwa' is almost round in form, very sweet and the most popular for landscape use. Fruit matures from October to May. The long fruiting period, fragrant blossoms and neat appearance are other assets.

Well located and protected from late frosts, a navel orange will produce a good crop of high quality fruit in Texas.

*Trifoliage orange—Deciduous and very cold hardy. Extremely thorny; leaves have three leaflets. Rather leggy tree that is very attractive when in bloom in spring. Hedge plantings make an impenetrable barrier. The fruit is fuzzy and inedible. The large, fragrant blossoms are very attractive. Trifoliate orange is most valuable as a rootstock for the upper Gulf Coast, but should not be used in areas with salty soils.

Much more information on these fruit types and varieties can be found in the chapter "Which Varieties Will You Choose?" beginning on page 40. Hardiness is discussed on page 30.

TRY CITRUS RELATIVES AND NEW VARIETIES

The same determination that has led commercial citrus growers in Texas to replant rather than give up after severe frosts has led some home fruit growers to search for hardier varieties of citrus. Varieties such as 'Nansho daidai' (or *Citrus taiwanica*), 'Eromocitrus', 'Pong Koa Honey' mandarin, and 'Lang Huang Kat' mandarin are regularly discussed by members of the Rare Fruit Council. For more information contact:

Dr. J. W. Nagle
114 Clear Lake Rd.
Kemah, TX 77565.

Growing Citrus Outside the Citrus Belt

It may seem contradictory, but citrus trees are both tender and tough. They are considered tender because most can be damaged by temperatures at freezing and below, and because even subtle aspects of climate will affect many of their characteristics. But citrus trees can tolerate soil, humidity and light conditions that many other plants cannot. Little has been written about growing citrus under inhospitable conditions, so we went to a number of people who had success and asked questions. Their comments are included here. Carl Withner of Bellingham, Washington told us of a personal experience that proved to him the durability of citrus trees: "I picked up my 'Otaheite' orange in New York's Chinatown. It was leafless and had been tossed into the street as trash. It not only survived, but is a healthy tree today."

Combined with the promise of fresh citrus fruits, this tenacity is the key to the great popularity of citrus as indoor plants.

AN INTRODUCTION TO INDOOR CITRUS

The citrus most commonly available from mail order sources is the 'Ponderosa', also known as 'American Wonder' lemon. Whichever name is used, this is actually a citron-lemon hybrid that produces exceptionally large fruit. It is very adaptable and can be grown successfully throughout the United States. For example:

"We must have had our 'Ponderosa' for at least 10

Clean, attractive foliage, fragrant blossoms and colorful fruit make citrus an aristocrat among indoor plants. Pictured at left is 'Dancy' tangerine.

years. It produced fruit in July and December for so many years that the family simply expected lemon pie on both July Fourth and Christmas. I've made as many as three pies from one fruit."—Mrs. Jean Mideke, Washington.

"My 'Ponderosa' tree is now three years old and three feet high. It has produced two huge lemons. One weighed 1½ pounds and the other weighed two pounds."—Mrs. Albert Unger, Bartley, Nebraska.

"As you might imagine, 'Ponderosa's' fruit size and productivity eliminate the need to purchase lemons at the local supermarket. You will be required to keep its ungainly growth in check. This is easily done by selective pruning of unnecessary branches in late May or early June when the plant is moved outdoors for the summer. Before it is returned indoors, the plant will produce great profusions of big, waxy, fragrant white flowers. When plants are outdoors, the bees take responsibility for pollination."—Richard Weir III, Cooperative Extension Agent, Nassau County, New York.

The 'Otaheite' orange is another easy-to-grow citrus that is well suited to indoor culture. Generally considered a dwarf form of the 'Rangpur' lime, the same plant is known as 'Tahiti' orange and sometimes 'Chinese New Year' orange. 'Tahiti' orange refers to the fact that plants were originally imported into this country from Tahiti around 1880. In fact, 'Otaheite' is a misspelling of the name of the city Otaite in Tahiti. Attractive and very sweet, deep orange-red fruits are most abundant in the time of the Chinese New Year.

Another popular indoor citrus is the 'Meyer' lemon, which is actually a lemon-orange hybrid. It is a versatile and adaptable citrus. The yellow fruits are used much as a regular lemon.

The 'Persian' lime or its California seedling, the 'Bearss' lime, produce less aromatic fruit than the 'Mexican' lime, but the trees are dense growing and very attractive.

The calamondin is actually a sour mandarin. It is one of the best known and easiest to grow. An excellent houseplant, it is usually covered by flowers and fruits.

These are some of the most common and easiest citrus to grow indoors. But of the more than 60 varieties of citrus described in this book, many are likely candidates. If you are interested, give any one of them a try.

SUCCESS WITH INDOOR CITRUS

There is no "right way" to grow any plant, and citrus is no exception. But the adaptability of citrus and the many variables of growing citrus indoor-outdoors allows success with a wide variety of cultural techniques. Many people keep their citrus trees indoors year-

Interior designer Lauretta Abbott of New Jersey appreciates the elegance of citrus indoors, and includes it in her designs.

Grapefruit tree (photos above) started by Dorothy Terhune of New Jersey, has grown indoors for 18 years.

round, treating them as any other houseplant. Some keep trees indoors in winter and move them outdoors in spring. A few keep citrus in a greenhouse. It is difficult to be specific about citrus culture in these varying situations. But here are the best general recommendations regarding soil, water, fertilizer, supplemental light, flowers, fruit and pollination, and the common problems you are likely to encounter.

Soil. Some plants will only grow well in a certain soil mix, but citrus accepts many soil types. Indoor growers have reported success using a wide variety of soils based on a commercial potting mix, peat moss or loam soil mixed with various amendments. In most cases commercial potting soils are best, used just as they come from the bag. These soils are lightweight and easy to handle. They are uniform in texture and free of disease organisms. Adding loam soil to such a mix destroys the predictability of pH and nutrient content, and increases the possibility of disease.

Standard container mixes are sold under a variety of trade names—Jiffy Mix, Jiffy Mix Plus, Metro Mix, Super Soil, Peat Lite, Baccto, Redi-Earth, Terra-Lite and so on.

The organic fraction may be sphagnum peat moss, pine bark, hardwood bark or fir bark. The inorganic fraction may be vermiculite, perlite or fine sand. Many of these mixes, except in the West, are patterned after the Cornell Peat-Lite Mix, which is 50% vermiculite, 50% sphagnum peat moss and mixed with the following for each cubic yard of soil: 5 pounds limestone, 1 to 2 pounds of superphosphate and 1 pound of calcium or potassium nitrate. Potting soils in the West are patterned after the University of California soil-less mixes. Typical examples follow.

To make one cubic yard of container mix:
    ¼ cubic yard sphagnum peat moss
    ¼ cubic yard propagation grade perlite
    ¼ cubic yard fine sand
    ¼ cubic yard ground bark (fir or pine)
    1½ pounds urea formaldehyde 38-0-0
    3 pounds single superphosphate 0-20-0
    1 pound potassium sulfate 0-0-50
    6 pounds calcium carbonate lime
    6 pounds dolomite lime
    1 pound iron sulfate

The urea formaldehyde 38-0-0 compensates for the nitrogen needed to decay the bark. You can substitute four pounds of blood meal or four pounds of hoof and horn meal for the same purpose.

The calcium carbonate and dolomite are finely ground limes and their amounts are typical for most areas. In areas of high calcium-magnesium bicarbonate waters, lower rates should be used. Check with your water department to find out the mineral composition of your water.

To make 2 bushels* of container mix:

1 bushel	Sphagnum peat moss
1 bushel	Horticultural vermiculite
10 Tablespoonsful	Ground limestone
5 Tablespoonsful	Single superphosphate
2 Tablespoonsful	Potassium nitrate
1 Teaspoonful	Iron chelate

*Approximately 1.2 cubic feet per bushel. One cubic yard equals 27 cubic feet or 22 bushels.

All of the ingredients of a potting mix need to be thoroughly blended. You can mix them easily by shoveling the components into a cone-shaped pile and then rebuilding it three times to ensure a thorough mix.

Water. Container citrus respond to water the same as any other container houseplant. You must maintain a balance between too much and too little water. Anne Glenn of Virginia has learned from this experience: "My trees must remain constantly moist. I have found this to be critical. At first I tended to let the trees dry out, using a moisture meter as a guide. But the trees suffered and some died. I began watering the survivors heavily and the results were dramatic. Now I wait until the top of the soil dries slightly and then I water thoroughly."

Keep in mind that most watering problems are related to the soil mix. There must be plenty of air space in the soil after water drains away. Generally, watering is less of a problem if you use a lightweight soil mix. Heavy and frequent watering is the privilege of gardeners using light well-drained soils. Mrs. Ofstun of Virginia adds: "Never allow the pots to stand in water. Good drainage is a must."

Fertilizer. Most every type of "complete" fertilizer—those that contain nitrogen, phosphorus and potash—can be used with good results. A steady, constant supply of nutrients, which can be provided by dilute amounts of a soluble fertilizer with each watering, is best. Slow release and the more common granule fertilizers can also be used.

Winter feeding and watering requirements will depend largely upon the tree's environment. If it is gradually exposed to cooler temperatures and allowed to become essentially dormant, little fertilizer or water will be necessary. However, in a warm house, nutrient needs will remain fairly constant. Fertilize according to the growth rate.

Supplemental light. Citrus trees adapt to relatively low light levels, but will grow faster and have more leaves when exposed to higher levels of light. Cecil C. Goodson of Indiana wrote, "The more light I give my plants and trees, the better results I get."

Standard fluorescent cool white, warm white, Gro-Lux fluorescent or combinations of fluorescent and incandescent light are typically used. The closer the standard tubes are to the leaves, without touching them, the better. Light intensity falls rapidly with increasing distance. "High output" or "very high output" fluorescent tubes, mercury vapor bulbs and high intensity discharge bulbs require special wiring, but can be kept four to six feet from the plant tops. Wonder-Lite is a commercially available bulb that combines the characteristics of mercury vapor and incandescent lights. It requires no special wiring and can be kept three to four feet from the plants.

Mrs. Jacobs of Shawnee Mission, Kansas, told us about growing plants under lights: "Growing fruit trees and other plants under lights has been a source of pleasure and a great hobby, especially during the winter months. Working around plants is extremely relaxing and we find it has a tranquilizing effect after a long day at the office."

Flowers and fruit. Many variables can affect flower and fruit formation on citrus trees grown outside of the citrus belt. Citrus trees will flower and fruit in almost every situation for nearly every gardener. However, there may be vast differences in frequency and quantity. Humidity and temperature are very important, but even more important is the care given by the gardener. Vigorous, carefully tended trees usually yield better on a more regular basis than slow growing trees receiving minimum care. Of the 100 growers of indoor citrus we contacted (Indoor Citrus Questionnaire, page 146), 85 have eaten fruit from one or more of their trees.

Citrus flowers alone may be reward enough for you. Mrs. Eleanor Plamondin of Oregon writes, "I think the fragrance is almost as enjoyable as the fruit, although I did originally buy the trees hoping they would fruit. Our attached greenhouse exchanges air with the house so we frequently have the fragrance of citrus blossoms indoors."

From Mrs. Sonya Falkenstern, New Jersey: "Even if I never got any fruit, I would grow these trees for their beautiful leaves, lovely flowers and heavenly scent. They always look good. With enough light, water and occasional fertilizer anyone can succeed with citrus."

Pollination. You'll have to help pollinate some citrus flowers if your trees are kept indoors or in a greenhouse all year. Use a small artist's brush, Q-Tips or pipe cleaners to transfer the pollen from one flower to another. If there are many flowers on the tree, a good shake may be enough to spread the pollen.

All citrus trees cannot self-pollinate. 'Clementine' mandarins will not set fruit without the pollen from a

'Dancy' or 'Kinnow' mandarin. Others, such as the 'Washington' navel, will produce fruit without any pollination. The chapter "Which Varieties Will You Choose?" includes specifics regarding pollination.

PROBLEMS

As with any plant in any situation, occasional problems do occur. Some citrus have trouble getting too much or too little water, but this is usually soil-related.

Humidity. The most significant recurring problem comes from periodic lack of humidity. There is a clear relationship between humidity and stress indoors. At 50 percent relative humidity and 75°F (24° C), plant stress is low. At the same temperature but with only 20 percent humidity, stress is great. The average relative humidity during the heating season is 20 percent or less. We know of homes that have been as low as 5 percent. This can create a great deal of stress when a citrus tree is brought indoors for winter. Adding a humidifier usually only raises humidity to about 30 percent. Mr. William Madonna of Rhode Island wrote, "By increasing humidity with a small humidifier (like one used in a sick room) you reduce shock caused by a change in environment from the out of doors. Citrus will not tolerate a dry atmosphere. My 'Marsh Seedless' did not bloom until I provided humidity in the winter months."

Bernard Diedrick of Kentucky adds that stress caused by lack of humidity can cause more severe results than simply lack of flowering. His words: "They seem to suffer from dry heat inside in winter with quite a bit of leaf drop and some branch dieback."

If the amount of water in the air stays the same, temperature will control the relative humidity. As temperatures drop, relative humidity increases. Also, lower temperatures slow the metabolic rate of the plant and so reduce its needs. Carl L. Withner, Bellingham, Washington, is aware of this fact: "In winter I keep the trees in a closed off bedroom so they are about 50°F or even colder at night. I feel a day-night temperature difference is important." It is. Try to have night temperatures about 10 degrees cooler than days. A significant day-to-night temperature difference will also promote deeper orange color on fruit.

Acclimating Citrus. The earliest orangerie citrus growers were familiar with the problems caused by inadequate humidity. It was discovered early on that citrus should be gradually acquainted with different climate conditions. In 1676, garden writer John Rea wrote, "You must on fairer days acquaint them again with the sun and air by degrees." Three modern citrus growers do just that:

"I move them first to my back porch, then under some trees until late September or about a week before I turn my house heat on."—Mrs. Selby Stepzinski, Illinois.

"I bring the trees in about mid September to avoid coinciding with the beginning of the heating season and because the house will start to get sunny again as the leaves fall from the deciduous trees."—Mrs. J. Alexis Burland, Pennsylvania.

"In fall the trees come into the house when indoor-outdoor temperatures are about equal and before the furnace starts up."—Mrs. Melville H. Pratt, Michigan.

The interior of the orangerie on the Alfred I. du Pont estate in Delaware, was built in 1909 specifically for growing citrus indoors.

In spring the citrus growing on the du Pont estate are taken outdoors. Included in their cold climate citrus collection are oranges, tangelos, lemons and calamondins.

Photography by David L. Thomas

Note that all three of these gardeners live where winters are long and summers are short. The trees need to be well adjusted if they are to pass their winter indoors in good shape. Gardeners in Oregon and Washington can worry less about this transition from outdoors to indoors.

Leaf burn. Leaf burn generally occurs during the transition from indoors to outdoors. Solutions are attained by using awareness and common sense:

"I move the trees into a shaded spot in the yard or on the back porch that receives less than two hours of direct sunlight a day. I leave the trees here for about 3 weeks in spring and then move them to a sunnier spot."—Robert Anderson, Ohio.

"I wait to move my plants outside until about mid May, or when the nights are warm. They are placed on a deck with filtered shade for most of the time. The plants do get direct morning sun. I keep them on the deck all summer."—Jefferson C. Hooper, Massachusetts.

"My trees are put in the garage for a month before the last frost date in spring. After that they are moved outside. I open the garage door for increasingly longer periods each day. If it is hot, I let the sun shine on them for only two or three hours, then shut the door. They are misted when the sun is not on them and the garage floor is kept wet."—Ernest O. Hoffman, Minnesota.

"Since my citrus trees receive strong sun all year, I

Kumquat tree (shown against rear wall) bears good quality fruit when grown in-ground in heated greenhouse—minimum winter temperature, 40° F.

have no transitional problems. It is just a simple matter of carrying them out of the greenhouse onto the patio."—Jerry T. Hamilton, Kentucky.

Pests. By far the most common troublesome pests of indoor citrus are scale and mites. Aphids, whiteflies, mealybugs and an occasional snail or slug are also troublemakers.

The most popular remedies are the insecticide malathion or simply washing the leaves. Ruth Strother of Indiana recommends the latter: "Handwashing of leaves with a washcloth, lukewarm water and Ivory suds is a real job, but it works!" Twice a year, as the plants are being moved, is a good time to wash them. This is especially important when the trees are coming indoors after attracting a variety of undesirables during a summer outside.

WE AGREE

It appears that citrus is not much more difficult to grow indoors than most of the popular houseplants. Of course the trees require some special attention to produce flowers and fruit indoors. We found a wide range of successful methods employed: technical to simple. There must be a wide margin of error within which one can balance the various needs of the trees. Echoing these sentiments, William Talbot of Connecticut writes, "Having seen citrus growing in Italy and Greece under the most difficult circumstances, as street trees, or in groves where they are exposed to cold, heat and are pruned excessively, I think one needn't handle them with kid gloves." And from Mrs. Melville H. Pratt of Michigan: "I'm a nurse and keep a large garden with not much time for all my indoor plants. But my citrus have been a great pleasure watching them bloom and bear. I plan to buy more. They have been very faithful and rewarding despite my lack of care."

INDOOR CITRUS QUESTIONNAIRE

With the Indoor Citrus Questionnaire we contacted 100 growers of indoor citrus. They live in 34 of the cold winter states of the United States, two are from Canada, one is from West Germany, and one is from Switzerland. Their names were obtained from two California nurseries that specialize in mail order citrus. Many of their words have been quoted in this section to illustrate certain points, and for that, we thank them all very much. The listings below show where the questionnaire respondents live, which citrus they grow and the situation in which they grow them.

SURVEY RESPONDENTS — WHERE THEY LIVE

States	Indoor/Outdoor	Indoor Only	Greenhouse Only	Total
MIDWEST				
Arkansas	1	0	0	1
Illinois	8	1	0	9
Indiana	0	0	1	1
Iowa	1	0	0	1
Kansas	3	0	0	3
Michigan	5	0	0	5
Minnesota	2	0	0	2
Missouri	0	1	0	1
Montana	2	0	0	2
Ohio	4	0	0	4
Oklahoma	1	0	2	3
South Dakota	1	0	0	1
EAST				
Connecticut	1	0	0	1
Maine	1	0	0	1
Maryland	2	0	0	2
Massachusetts	4	2	0	6
New Hampshire	1	1	0	2
New Jersey	1	0	0	1
New York	4	1	1	6
Pennsylvania	5	1	4	10
SOUTH				
Alabama	1	0	0	1
Georgia	1	0	1	2
Kentucky	2	0	0	2
North Carolina	1	0	0	1
Tennessee	4	1	0	5
Virginia	2	1	0	3
West Virginia	0	1	0	1
WEST				
Colorado	2	0	2	4
Hawaii	1	0	0	1
Nevada	0	0	1	1
Oregon	1	0	2	3
Washington	10	2	1	13
WEST GERMANY	0	0	1	1
CANADA	0	0	2	2
Numbers totaled	70	12	18	100

SURVEY RESPONDENTS — THE CITRUS THEY GROW

Lemons:	
'Eureka' or 'Lisbon'	20
'Meyer'	36
'Ponderosa'	34
Limes:	
'Bearss' (or 'Persian')	40
'Mexican'	4
Mandarins:	
'Clementine'	6
'Dancy'	13
'Kara'	5
'Kinnow'	12
Satsuma	9
Oranges:	
navels ('Washington' or 'Robertson')	39
'Valencia'	21
blood oranges	10
Grapefruit:	
'Redblush'	16
'Marsh Seedless'	7
Calamondins	14

MAIL ORDER CITRUS

Mail order citrus from California nurseries is shipped bare root wrapped in damp peat moss. The nursery process and what you can expect in the mail is pictured on these pages. Citrus trees from most other mail order nurseries are actually grown in and shipped from Florida. Unless otherwise noted, they are rooted cuttings and will be shipped in 2-1/2 or 3-1/2 inch plastic pots. A 'Ponderosa' lemon and 'Otaheite' orange are shown as supplied from such a nursery. See photos, right.

Citrus shipped bare root should be planted immediately. If this is not possible, keep the roots moist and store the tree in a cool place out of direct sunlight. **Note:** In some states where citrus is a commercial crop, citrus plants grown in other states cannot be shipped into the state. Variables do exist. Check with your mail order source.

SOURCES

Alberts & Merkel
2210 S. Federal Highway
Boynton Beach, FL 33435
Tropical plant specialists that also supply many varieties of citrus. They ship in 6, 8 or 10-inch containers. Catalog $1.

Banana Tree
715 N. Hampton Street
Easton, PA 18042
Specialty catalog, 5-1/2 by 8-1/2 inches, 16 pages. Send 25¢ for postage and handling. Source for trifoliate orange and calamondin. Also offers a wide range of rare seeds, bulbs and plants, including 33 varieties of bananas.

Burgess Seed & Plant Co.
905 Four Seasons Road
Bloomington, IL 61701
Two separate catalogs are available, one for indoor gardeners and one for outdoor gardeners. Both are 32 pages, 8-1/2 by 11 inches, and cost $1 each. As indoor plants they offer the 'Otaheite' orange, 'Persian' lime, 'Ponderosa' lemon, 'Paradise' grapefruit (same as 'Redblush') and 'Dancy' tangerine. They cannot ship plants into California, Arizona, Texas or Florida.

Four Winds Growers
42186 Palm Avenue
Box 3538
Fremont, CA 94538
Wholesale grower of true dwarf citrus for western nurseries. Mail order available to other states. Two-gallon and five gallon sizes shipped bare root. Over

Citrus purchased from California mail order catalogs are shipped bare root. Grafted varieties on dwarf rootstocks are taken from 2 gallon and 5 gallon containers. All the soil is removed. Roots are packed in moist bark shavings and plastic. Trees are placed in shipping boxes and sent out. If you can not plant bare root citrus immediately, keep roots moist and place tree in a cool location away from direct sun.

Florida-grown mail order citrus trees are shipped in 2½" and 3½" pots. 'Ponderosa' lemon and 'Otaheite' orange above.

30 varieties available. All orders are accompanied by the pamphlet *How to Grow Citrus* and a catalog. Pamphlet costs 25¢ without order.

Gurney Seed & Nursery Co.
Yankton, SD 57079

Free 66 page, 16 by 20-inch, general nursery catalog. Eight pages of fruits and berries, and one page of indoor plants. In addition to lemons, oranges and indoor plants, they offer a "micro-budded" dwarf grapefruit.

Kelly Bros. Nursery Inc.
Dansville, NY 14437

General nursery catalog, 72 pages, 8-1/2 by 11 inches. Specialties are fruits and berries. One page of houseplants. They offer a dwarf orange ('Otaheite') and the 'American Wonder' lemon ('Ponderosa').

Lifetime Nursery Products
1866 Sheridan Road
Highland Park, Il 60035

Catalog is 28 pages, 5-1/2 by 8-1/2 inches, specializes in amaryllis and unusual plants. Citrus are available in different sizes. In one gallon containers: grapefruit, lemon, lime and kumquat. In smaller containers: oranges, lemon and lime.

Pacific Tree Farms
4301 Lynnwood Drive
Chula Vista, CA 92010

Specialty catalog, 5-1/2 by 8-1/2 inches, 16 pages. Carries over 50 varieties of citrus. Also offers an interesting selection of trees, rare fruit and garden supplies.

Van Bourgondiens
245 Farmingdale Road
P.O. Box A
Babylon, NY 11702

Catalog is 52 pages, 8-1/2 by 11 inches, primarily concerned with bulbs. There are four pages of fruits and berries and two pages of houseplants. Catalog includes: 'Ponderosa' lemon, 'Persian' lime and the 'Mitis' dwarf orange (same as calamondin).

5 Landscaping with Citrus and Their Relatives

Few gardeners or landscapers have had the opportunity to observe the many kinds of citrus and citrus relatives for extended periods of time. For over 30 years horticulturist W.P. Bitters of the University of California at Riverside has observed the many varieties and species with an eye to ornamental possibilities. The following article was first published in *LASCA Leaves,* a publication of the Los Angeles State and County Arboretum. It is acknowledged as the best treatment of the use citrus in the landscape. We have omitted some of his words that pertain to subjects covered elsewhere in this book. We quote:

"A discussion of citrus for ornamental purposes would not be complete without considering the history and distribution of *Citrus* species. It was the ornamental aspect of the trees and fruit, and the pride and distinction that accompanied possession of such trees that were more important in the spread of *Citrus* than the taste of the fruit.

"The conquest of Alexander the Great brought the citron to Greece. Later, conquests by Rome extended it to the shores of Italy. Both Greeks and Romans recognized its decorative value. Many mosaics, murals and sculptures that survive from their time include the citron.

"There are many legends concerning citrus in mythology. (Some are more fully detailed in the first pages of this book. If you read them, notice that most relate to the beauty of the tree or fruit, not the fruit's taste.) One tells of the wedding of Jupiter and Juno and how a tree bearing "golden fruit" suddenly appeared. The gods were so proud and protective of the tree they sent it to the Isles of Hesperides and left Atlas as guard. Another myth tells how a Grecian maiden

Once part of a productive orchard, these grapefruit now grace the landscape of this Scottsdale, Arizona home. White painted trunks are attractive and protected from sunburn. Use a standard "outdoor" water base paint.

was tricked into marriage when her suitor distracted her with "golden apples."

"There are many such legends and how or when they appeared is unknown. But keep in mind that only inedible citron was known at this time. Its attraction by color and fragrance alone was apparently enough to generate myths.

"The Moors (Arabs) spread the sour orange throughout northern Africa and into Spain. They recognized the ornamental value of the sour orange and incorporated it into their planting designs. Beautiful mosques were built in Cordoba, Granada, Seville and many other Spanish cities. Supposedly, the largest and most magnificient in all Islam was the Mosque of the Omyyads, built at Cordoba in 976 A.D. In connection with it was the *Patio de los naranjos*, or courtyard of the oranges. Sixteen rows of sour orange trees, each running up to one of the arched entrances of the mosque, were oriented so the axis of the orange rows were in direct line with the pillars inside. This is one of the best examples of the symmetry and integration of landscape and architectural design which is so characteristic of moslem architecture.

"Orangeries, or buildings for growing citrus, spread throughout Europe during the Middle Ages and the Renaissance as a result of the Crusades. The first book devoted entirely to citrus was printed in 1646. Written by Ferrari and entitled *Hesperides*, it contained many drawings of orangeries. The most elaborate was built by Louis XIV at Versailles, France. He used his orangerie to have potted ornamental citrus and citrus blossom fragrance throughout his castle year-round.

This Phoenix, Arizona landscape is based in part on the traditional look of a citrus orchard in the desert. Palms were frequently planted to shade the citrus trees from the intense sunlight.

CITRUS IN THE LANDSCAPE

"In planting design, citrus may be used for hedges, basal plantings, shade, corner accent, framing, individual specimens and backgrounds. Following are the groups of citrus and citrus relatives best adapted to landscape use.

'Bouquet' Sour Oranges, The 'Bouquet' sour orange (or 'Bouquet de Fleurs') is occasionally grown in both California and Florida as an ornamental. The variety has been often confused with and mistakenly called the 'Bergamot' orange. Specimens rarely grow over 8 to 10 feet high and are somewhat spreading with dense foliage which tends to cluster because of the short internodes (distance between leaves). An extensive hedge of this citrus may be found at the Citrus

Navel orange can be a landscape focal point. The attractive qualities of this one are emphasized by the border of gazanias.

For some homes a miniature citrus orchard is a practical, water-saving alternative to a lawn. Stone mulch between trees is attractive, prevents most weed growth and reflects heat and light into the trees.

Mature 'Bouquet' sour oranges rarely exceed 10 feet in height and are relatively wide spreading. Use them as you would pittosporum—as a shrub or small tree. Outstanding fragrance is an added bonus.

This hedge of 'Bouquet' sour oranges is over 40 years old and requires a minimum of pruning. The location is the Citrus Experiment Station at Riverside, California.

Flowers of 'Bouquet' sour orange: They appear similar to other citrus flowers, but their fragrance is by far the most delicate and penetrating of them all.

Experiment Station, Riverside, California. The hedge is now 40 years old and still healthy and beautiful. It requires a minimum of pruning. As a shrub it can be used in situations similar to those in which pittosporum is commonly grown. The tree can also be used for framing, corner accent, an individual specimen, or even backgrounds. This variety is extensively grown in southern France where the flowers are used in the manufacture of perfume.

'Chinotto' and 'Myrtifolia' Oranges. Both apparently variants of the sour orange, they are similar, but with notable differences. The 'Chinotto' has a slightly broader leaf than the 'Myrtifolia' and a more open

Leaves of 'Bouquet' sour orange: An attractive shiny green like most citrus, but in dense clusters. Hence its virtues as a hedge.

type growth. The 'Chinotto' annually bears a very heavy crop of deep orange fruits. They persist throughout the season, superimposed on the dense, fine-textured foliage, making this citrus *ne plus ultra* of the natural dwarf selections. 'Chinotto' juice is usable for making a refreshing drink. Some strains are seedless and are prized for the preparation of candied oranges, jellies, preserves and other similar products of distinctive character.

"The 'Myrtifolia' is a small dwarf tree or shrub with thornless branches and very small, closely set leaves. The leaves are only about one-third as large as a standard sour orange and their close spacing makes a rosette-type growth. The foliage is thus very dense and compact and the tree is very symmetrical. Growth is generally slightly columnar to conical and mature trees 25 years old at Riverside are only about 10 feet in height. The tree has a prolific bloom but, in contrast to the 'Chinotto', sets very few fruit. The 'Myrtifolia' is similar in appearance to pittosporum and can be used for hedge, backgrounds, individual specimens or corner accent.

A willowleaf form of the sour orange, one of the many varieties of this long-favored landscape citrus.

"**Sour Orange.** The sour orange (*Citrus aurantium*) has many varieties. Some have long, lanceolate, willow-like leaves. Others may have variegated (sectoral) fruits of contrasting colors. The sour orange was used extensively in the landscaping of the mosques, courtyards and public buildings of the Arabs. In recent years, its principal value has been as a rootstock. Its fruit is used in the making of marmalades and also make an excellent pie. However, because of its very attractive, brightly colored, reddish orange fruits, and its lush, dense, dark green foliage, the sour orange still has considerable merit as an ornamental. In some cities the sour orange is used for streetside planting. Examples may be seen in Tempe, Arizona, or near Litchfield Park, Arizona. Scripps College, Claremont, California, has effectively used sour oranges in patios, courtyards and drives. It may be best used as a shade tree, individual specimen, background or corner ac-

The sculptured trunks on these sour orange trees adds interest to the entrance walkway.

cent. Sour oranges probably should be budded. Seedlings may be grown, but they tend to be more columnar and thorny.

"Citron. The citron is one of the oldest of the citrus fruits and certainly the one best known to Mediterranean countries during ancient and medieval times. It is not commonly grown in the United States because the trees are fairly sensitive to frost, but no more so than the limes. Many areas in California certainly possess a suitable microclimate. The trees are generally grown as cuttings and are dwarf-size. They tend to be short-lived. The citron blooms almost throughout the year. The leaves resemble a lemon, but they are generally coarser and more leathery. Tree form is generally open or sparsely foliated and the fruits are large, averaging five to seven inches in length and 3 to 5 inches in width. Fruit is principally used for its peel. Citron rind is generally more than an inch thick and is used in the production of candied peel which is extensively used in the flavoring of confections and cakes. The pulp is not eaten. "Citron water" may also be manufactured from the fruit and may be used in flavoring of liqueurs and vermouth and for medicinal purposes. Because of the thick rind, the fruits keep for

This streetside planting demonstrates one of the best uses of the sour orange. The bright orange fruit is too bitter to be eaten, but decorate the trees almost continuously.

a long time and are very attractive when used in table displays.

"The fingered citron ('Buddha's Hand') is not known to be used in California as an ornamental, although it has been introduced under the University's new citrus importation program. Fruits of the fingered citron are highly esteemed for their fragrance and are used extensively by the Chinese and Japanese for perfuming rooms and clothing. Since the tree is dwarf, it can be grown as a potted plant or as an individual specimen.

" **'Rangpur' lime.** The 'Rangpur' lime is an acid lime, sometimes called a red lime. It has at times been confused with the 'Otaheite' orange, discussed next, and substituted for it. Differences are the 'Otaheite' is seedless and non-acid and the 'Rangpur' is both seedy and acid.

"The 'Rangpur' lime is easily rooted from cuttings and is frequently grown as a shrub or a dwarf. It has a bush-type growth and may eventually reach a height of 15 or 20 feet. Apparently there are strains, so fruit color and shape vary somewhat. Generally the fruit color ranges from yellow-orange to reddish orange. The flesh is a deep orange.

"The fruit could easily be mistaken for a tangerine and this confusion has probably led to its limited use. The juice is orange in color, has an excellent lime-like taste, and could easily be substituted for limeade. The chief use is as a dwarf in the early years, but older trees grow large and make an attractive individual specimen or background.

" **'Otaheite' orange.** The 'Otaheite' orange is somewhat of a misnomer since it is not an orange but is probably nearer to being an acidless lime or a hybrid thereof. It is a natural dwarf and is easily grown from cuttings. The variety is nearly thornless and spreading in growth. The flower buds are tinted with purple and the new leaves are a deep purple. The fruits are chiefly spherical in shape and range in size up to about 1¾ inches in diameter, thus smaller than the 'Rangpur' lime. The fruits have orange-colored flesh, are seedless or have abortive seeds, have the odor of a lime, but are flat and insipid in flavor. The 'Otaheite' is frequently grown as a potted ornamental. They are especially attractive at Christmas time since plants a foot or so high may by carrying up to a dozen fruit in addition to many new flowers. In fact, this would appear to be its most practical use.

The sour orange can assume many shapes, as dictated by your pruning shears.

The extra-close one foot spacing of these sour oranges made an almost instant hedge.

More common today as dwarf or indoor citrus, these full size, columnar calamondins are spectacular as landscape specimens. Each is carrying hundreds of fruit.

One of the most landscape adaptable citrus relatives is the Chinese box orange. Naturally compact, it can be pruned to make a dense hedge.

"Calamondin. The calamondin tends to grow tall and columnar, and is very shapely in appearance. As an ornamental it has been widely used in Florida and to a lesser extent in Arizona and California. The calamondin is very cold-resistant and easy to grow. The tree is a prolific bearer and may have hundreds of fruit at any one time. The Chinese name for it is *Sechi Chief*, meaning four seasons. The name is appropriate because the fruit is produced year-round. The quantity of fruit makes the tree spectacular in appearance. There may be hundreds on a mature tree. The fruit is orange colored, slightly flattened in shape, and about ¾ to 1½ inches in diameter. The fruit makes excellent marmalade and the acid juice makes a very presentable drink when sweetened with sugar. Because the tree is tall and columnar, it may be best used as a background tree, shade tree, individual specimen or corner accent. On a dwarfing rootstock it makes an excellent pot variety. A variegated selection is also available.

"Chinese box orange. A citrus relative, it belongs to the genus *Severinia*. The common name derives from its similarity to the common box, *Buxus sempervirens*. There are many forms that differ in height, character of growth, size and shape of leaves, thorniness, etc. Generally they are low shrubs or dwarf trees with somewhat rounded leathery leaves about 1½ inches long attached by short stalks to the twigs. The internodes are short, so the leaves are densely crowded on the branches, making growth very compact. The fruit is small (about the size of a pea), round, dark blue (nearly black), and filled with one or two large seeds. The plant may be propagated by cuttings, seed or budding. It is graft-compatible with citrus. Trees 26 years of age are growing on citrus rootstock at the Citrus Experiment Station at Riverside. The chief value of this species is its foliage. The plant is symmetrical in shape and the beautiful, dense, vivid-green leaves make the small leaf varieties an excellent substitute for the common box as a hedge. It takes pruning well, but is compact enough to not require pruning. In addition to being quite cold tolerant, it also appears to have excellent resistance to pests.

"Fingerlime. Sometimes called *Australian wild lime*, this is a near relative of citrus in the genus *Microcitrus*. Most grow into tall shrubs or small trees and make handsome ornamentals. The seedlings are very spiny. The leaves are probably the smallest of any of the true citrus or citrus relatives. They average about 1 inch long and ¼ to ½ inch wide. The habit of growth is such that young trees have the appearance of a very dwarfed fir tree. New growth is long, pendant and purplish in color. The flowers are small, pink and

The bright orange fruit of the kumquat adds considerable interest to this planting of evergreen shrubs. Kumquats are entirely edible—the skin is sweet and the flesh tart. This is the round-fruited variety 'Meiwa'.

fairly attractive. Fruits are about 3 inches in length and ¾ inch in diameter. They are aromatic and could probably be used for making pickles or preserves. Because most of these varieties are quite vigorous, they could be used for hedges or windbreaks. Pruning the lower branches produces an attractive shade tree useful as an individual tree or a background specimen. Propagate by budding onto citrus rootstock, cuttings or seeds.

"**Kumquat.** These are well-known citrus relatives of the genus *Fortunella*. [They are described at length elsewhere, but here we would like to draw attention to their considerable ornamental value.] In China, bearing potted plants are sometimes placed on the table during dinner. Guests can then pick and eat the fruit between courses. In the landscape, kumquats make a fine hedge. They are very resistant to cold and can be grown in many areas where other citrus will not survive.

"**Limeberry.** The limeberry (*Triphasia trifolia*) is a citrus relative widely grown as an ornamental shrub in tropical and subtropical areas. It has a compound, trifoliate leaf and fragrant white flowers. Fruits are dull red and about ½ inch in diameter. The foliage is an attractive, shiny, dark green. The plant makes a small, round-topped, shrub, suitable for dooryard plantings or potted specimens. It has become naturalized to some areas. It is now growing at Riverside and seems very worthy of consideration for ornamental use.

"**Trifoliate orange.** The Trifoliate orange (*Poncirus trifoliata*) is widely propagated in the east as far north as New Jersey because it is so tolerant of cold. It has a trifoliate leaf, is extremely thorny, and is deciduous. The flowers are very conspicuous in the spring since they precede the leaves. The fruit is about the size of a golf ball, very seedy, and inedible. Plant is frequently used for hedges but appears to have no place in mild winter areas where other citrus can be used instead.

A mature 'Sinton' citrangequat: Fruit is not edible, but is highly ornamental through winter and spring.

Orange jessamine: Flowers are fragrant and leaves are shiny.

Murraya flowers are large and yield edible fruit.

" **'Sinton' Citrangequat.** This is a common and very attractive ornamental in Florida and has recently been introduced to California. It is a "trigeneric" (three genera) hybrid of trifoliate orange, sweet orange and kumquat. The inedible fruits are deep reddish orange and conspicuous throughout the winter and spring. The trees are very hardy and could be grown out of the range of normal citrus.

" **Orange jessamine.** This citrus relative of the genus *Murraya* is perfectly adapted for ornamental use. It has a compound leaf, finer in texture than the Wampee (see below), very similar to a wisteria leaf but more delicate. The small leaflets are shiny dark green. The twigs are thin, spineless and flexible, providing a dense but pendant type of foliage. The flowers are white, fairly large and conspicuous, hanging in large clusters at the tips of the twigs or side branches. These blooms are exceptionally fragrant and reminiscent of jasmine. The tree is practically an everbloomer at Riverside and its delightful fragrance graces the air throughout the season. The *Murraya* is fairly fruitful and the species at Riverside sets a small red fruit about ½ inch long and ¼ inch wide. The variety question is a little confusing with *Murraya*. Specimens observed in Florida were more fruitful but less vigorous than those observed in California. The red fruits on a contrasting green background coupled with the attractive and fragrant bloom make the variety the acme of citrus ornamentals. At Riverside, *Murraya* withstands cold as well as commercial citrus and is practically pest free. The plant takes pruning well. In the landscape, use it for basal planting the same as cotoneaster or abelia—as a corner accent plant, as an individual specimen or as a background plant. The foliage provides excellent greens for floral arrangements.

" **Wampee.** The Wampee (*Clausena lansium*) is not a true citrus but is a relative. It grows as a tree and may attain a height of 20 feet or more. The leaves are compound, somewhat similar to those of a native black walnut in size and shape. The leaves are rougher in texture than most citrus and more yellow-green in color. The nodes are very close together giving rosette type of growth to the twig and consequently providing a very dense foliage cover. The flower is unique in that it is large and occurs as a cluster at the ends of the branches much like a lilac flower. At Riverside, fruits mature in early fall and are in open clusters much like a thin bunch of grapes. The fruits are yellowish brown to russet-brown in color, the size of a quarter and are almost round. They are edible and highly prized by Orientals. The peel is very thin and the pulp very tender, similar to a grape or litchi. Many are one-seeded and some only have abortive seeds. While the

Wampee can be budded on citrus rootstock, it may be better to grow it from a cutting or seed. It would make an excellent shade tree, background tree or specimen tree. It may be too coarse textured for other purposes. The Wampee appears much freer of insect pests than standard citrus varieties.

"The ornamental qualities of citrus are well summed up by the French writer Gallesio who in 1811 wrote: 'Of all the plants spread by nature upon the surface of the globe, there are none more beautiful than those we know under the names of citron, lemon and orange trees, which botanists have included under the technical and generic name *Citrus*. These charming trees are both useful and ornamental. No others equal them in beauty of leaf, delightful odor of flowers or splendor and taste of fruit. No other plant supplies delicious confection, agreeable seasonings, perfumes, essences, syrups and the valuable aides so useful to colorers. In a word, these trees charm the eye, satisfy the smell, gratify the taste, serving both luxury and art and presenting to astonished man a union of all delights. These brilliant qualities have made the citrus a favorite in all countries.' "

LANDSCAPE BASICS

If you have gone straight through to this point you have read a great deal about landscaping with citrus. In the following discussion we offer our personal experiences. It is our aim to bring the subject down to earth and make it easier for you to select your plants at the nursery.

THE CITRUS HEDGE

First, decide on the desired height of the hedge, then select the proper variety or varieties to fit the need. There are a number of selections that tolerate rather severe pruning. These include 'Eureka' and 'Lisbon' lemons, 'Meyer' lemon and 'Rangpur' lime. Most other types tolerate some crowding and shaping as long as they are pruned to enhance their natural shape and size. Dwarf navel oranges, grapefruit, 'Valencia' and other sweet oranges can be pruned to form a six to eight-foot hedge. A six-foot or lower citrus hedge can be made with mandarins. Kumquats make a neat three to four-foot hedge. 'Rangpur' lime and 'Meyer' lemon can be kept low, tolerating rather severe pruning.

The true lemons, 'Eureka' and 'Lisbon,' can tolerate rigorous pruning, but will sucker excessively if kept lower than five or six feet. On the other hand, 'Chinotto' and 'Bouquet de Fleurs' make very nice hedges with minimal pruning.

Regarding pruning, keep in mind that with very few exceptions, *the heavier the pruning the fewer the fruit.*

Sour oranges are pruned to make columns and a hedge.

A row of grapefruit completely hides the block wall behind it and produces a substantial harvest.

Orange jessamine: Flowers are fragrant and leaves are shiny dark green.

Citrus can be used for framing.

CITRUS SHRUBS AND CITRUS IN CONTAINERS

Dwarf forms of kumquats and kumquat hybrids such as limequats and orangequats can be used as basal or foundation plants. With some pruning they are easily kept as low as three feet.

By using two trees of the same variety, or two trees with similar growth habits to make a pair, citrus is very effective for framing. They may be suitable for entry or walkways whether planted in the ground or in containers. The size of the container is dictated by the variety. The navel and 'Valencia' oranges are best served by a 16-inch or larger container. The espaliered Satsuma in the photograph (far right) grows vigorously in a 10- by 16-inch container. Half-barrels, clay pots, wooden boxes or tubs also work well.

A few citrus trees can create your own mini-orchard.

Citrus varieties border a lawn, but have lower water needs than the lawn.

Heavy clay containers dress up these trees and give them an air of Mediterranean elegance.

CITRUS ESPALIERS

Espaliers are trees trained to grow flat against a wall or trellis by selective pruning. They are often trained into formal geometric design called *cordon*, *tier*, *candelabra* and *fan* depending on pattern. Informal training is much safer and easier than the stiff, strict, geometric patterns. Sunburn is less a problem with a more dense tree, and less frequent attention to pruning and tying is required of a gardener.

Espaliers save space and can well utilize the warm microclimate of a south-facing wall. Using this method, citrus can be grown in cooler regions. Most citrus can be espaliered, but some are better suited than others. Best are the true lemons, 'Summernavel' orange, 'Ponderosa' lemon, 'Tarocco' blood orange, the citron and 'Chandler' pummelo.

An arbor of lemons becomes hedge as seen from the outside (above) and shaded walkway from the inside (below).

Three espalier styles: Top, informal, but with a guiding framework; middle, formal—try this only with lemons; bottom, most casual, a selectively pruned navel orange.

Valencia orange   Page mandarin   Meyer lemon   Marsh Seedless grapefruit   Bearss lime   Moro blood orange

Washington navel orange **Fairchild mandarin**

6 Enjoying Citrus

This chapter is about rewards. Whether you grow your own citrus or buy it at the supermarket, the reward is the fresh, clean taste of citrus. From the fragrant, brightly colored fruit on the kitchen table to refreshing, fresh squeezed juice to delicious combinations with other foods, citrus is one of the most versatile and nutritious fruits.

We offer only a few recipes on these pages. Recipes that use citrus could make another whole book. However, we give you a start on where to look for the recipes and hints on how to use citrus in your everyday kitchen activities. Plus, there are tips on citrus storage, juicing, slicing and peeling.

For maximum enjoyment of citrus, you must have the best quality fruit.

The best way to tell if fruit on the tree is ripe is to pick one and taste it. If it is not sweet enough, wait a while longer. If it is dry, it has been on the tree too long.

Use pruning shears to harvest the fruit rather than pulling it off. Clip the stem near the fruit with a smooth cut. Try not to damage the fruit because cuts and bruises speed deterioration after harvest.

If you are selecting citrus at the market, look for firm, heavy fruit. The rind should be undamaged and have a good color. Puffiness of the rind is usually a sign of overripeness.

Sweet and tart, pale or dark, fresh citrus juice is a most pleasurable citrus experience.

'Moro' blood oranges yield the deepest red juice.

Juicing Fresh Citrus

Squeezing the juice from fresh citrus got its start as an American breakfast tradition when an advertisement entitled "Drink an Orange" appeared in the *Saturday Evening Post* magazine on February 19, 1916. Because of expanded orange production in California, the Sunkist Growers Association needed to increase public demand for their fruit. Through the *Saturday Evening Post* advertisement, they offered a heavy glass juice extractor for 10¢. With some effort it was used to extract juice from all sizes of citrus, from oranges to lemons.

Since the introduction of the first manufactured home juicer in 1916, the proliferation of equipment for preparing, juicing or grating citrus parallels its rise in popularity. Many of the items are pictured on these pages and throughout this book. Most can be found in the housewares department of department stores or specialty kitchen shops. They include a multitude of juicers in every shape and size, graters, peelers, scrapers and slicers—even thin cloth sacs so the seeds won't escape when you're squeezing lemons.

Today, juice from oranges, lemons, grapefruit, mandarins and limes is the most popular fruit drink in America. Industries have been founded on its preparation and packaging so that it can cheaply and easily reach any household. Still, there is something special about freshly squeezed juice.

TIPS ON SQUEEZING FRESH JUICE

• If you're buying fruit at the supermarket, pick fruit that is heavy for its size. They are juiciest.
• Fruit held at room temperature will yield more juice than chilled fruit.
• Before cutting for juicing, roll the fruit between the palm of your hand and a hard surface. This will burst some of the vesicles and allow the juice to be more easily released.
• To "freshen up" the flavor of processed juice, add the juice of a few freshly squeezed fruits.
• Don't be afraid to mix and blend the juices from several varieties. The results can be a pleasant surprise.

TWOSOME ORANGE-CRAN NOG (left)

1 cup crushed ice	¼ cup light cream or
½ cup fresh squeezed	half & half
orange juice	2 eggs
½ cup cranberry juice	2 tablespoons sugar
cocktail	2 orange cartwheels

In blender, combine all ingredients except orange cartwheel slices; blend until smooth. Pour into two 10-ounce glasses. Garnish each with orange cartwheel slice. Makes two 8-ounce servings.

A few of the many tools designed to extract citrus juices: Look for them in department stores and kitchen speciality shops.

JUICE FOR YOU

Oranges—More people probably juice oranges than any other citrus fruit. 'Valencia' is the traditional juice variety but most produce tasty juice.

Although varieties differ in juice content, expect about one cup of juice from three to four medium-size oranges.

Mandarins and mandarin hybrids—Mandarins offer a wide array of distinctive and very delicious flavors. The same is true for their hybrids, tangelos and tangors. You might want to use their grated peel in recipes. They yield exotic flavors.

The fruit size and juice content of these fruits varies too much to estimate the amount of juice in a standard-

Instant energy orange nog: Combine fresh squeezed orange juice, one egg, one tablespoon honey or sugar with two or three ice cubes.

Enjoying citrus **165**

FRESH LEMON IDEAS

1. Add fresh lemon juice to sliced apples, bananas, avocado and pears to prevent discoloring. (Photo above.)
2. Substitute lemon juice for seasonings.
3. Diet drinks taste better with a squeeze of lemon.
4. Squeeze lemon into a glass of water for a refreshing pick-me-up.
5. To make your own buttermilk add enough milk to 1 tablespoon lemon juice to make 1 cup.
6. Squeeze a lemon into tomato juice.
7. Roll lemon; poke hole in end with tooth pick; squeeze and out comes juice; replace pick for storing.
8. Squeeze juice over plain jello or ice cream.
9. Add fresh juice and a slice or two of fresh lemon to lemonade concentrate for a fresher taste and look.
10. After squeezing juice over fish save rind to scrub odors from frying pan.
11. Keep white vegetables white by adding lemon to the cooking water.
12. Serve hot lemonade in cold weather and cold lemonade in hot weather.
13. Substitute 1 teaspoon lemon juice for cream of tartar in a 3 egg meringue.
14. Bring lemon to room temperature and roll before squeezing to yield more juice.
15. Lemon Butter: combine 1 tablespoon chopped parsley, 1 teaspoon grated lemon peel, 1 tablespoon lemon juice with ⅓ cup butter.
16. Combine 1 tablespoon lemon juice with ¼ cup melted butter. Pour over 2 quarts popped popcorn.
17. It's easier to grate peel from an unsqueezed lemon.
18. Use lemons and salt to clean copper.
19. To remove scratches from wooden furniture, rub with mix of equal parts of lemon juice and salad oil.
20. Lemon juice cleans and adds a shine and sparkle to windows, glasses and decanters.
21. For discolored aluminum pots, boil sliced lemon and water to renew shine.
22. Place cut lemon in refrigerator for a deodorizer.
23. Freeze strips of lemon peel in ice cubes.
24. Keep cheese soft by rubbing with lemon and wrapping in damp cloth.
25. Toss used lemons into garbage disposal to keep it smelling fresh.

size fruit. Remember, mandarins and mandarin hybrids should always be held under refrigeration.

Lemons – Lemons are the most versatile citrus fruit. Most great chefs keep some lemons within easy reach. Uses for lemons seem endless and range from simple lemonade, to removing the smell of garlic from your hands, to flavoring fish. Many people use lemon juice instead of vinegar in salad dressings because vinegar masks the flavor of wine which may accompany a meal. A listing of some of the many uses of lemons can be found on this page. Expect about one cup of juice from six medium-size lemons.

Grapefruit – Grapefruit are becoming increasingly popular for fresh juice especially as a mix with alcoholic beverages. A medium-size grapefruit should yield at least ⅔ cup of juice.

Limes – 'Mexican' lime has a distinctive aroma but the 'Persian' or 'Bearss' has the most juice. Expect one cup of juice from six 'Bearss' limes. It will take eight or nine 'Mexican' limes to get the same amount of juice.

Fish with fresh-squeezed lemon juice is one of the oldest food combinations. Lemon delight fish fillets are a gourmet version of the traditional idea.

Fresh lemonade with home grown 'Eureka' lemons. Mix 2 cups of fresh lemon juice with 2 cups of sugar and 1½ cups of water. Add 2 teaspoons of grated lemon peel.

STORING FRESH JUICE

The fresh juice of most citrus varieties can be stored at 40° to 50° F (4°-10° C) in the refrigerator for 24 to 36 hours. After that it will lose its flavor. Juice from navel oranges is an exception; it usually becomes bitter after about four hours under refrigeration.

Keep chilled fresh juice covered in the refrigerator or it may pick up flavors from other foods. Before serving, stir it up as the pulp usually settles to the bottom.

The juice from lemons and 'Valencia' oranges can be frozen for up to four months. A convenient method is to freeze it in ice trays.

Limeade is even more refreshing than lemonade on the hottest days. Make it the same as lemonade but with slightly more sweetening to suit your taste.

Use The Whole Fruit

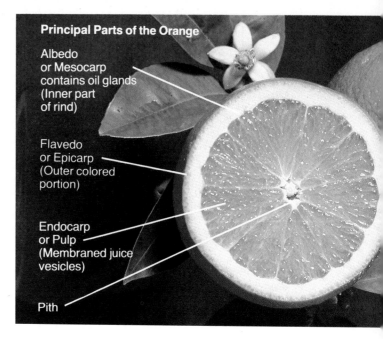

Principal Parts of the Orange

Albedo or Mesocarp contains oil glands (Inner part of rind)

Flavedo or Epicarp (Outer colored portion)

Endocarp or Pulp (Membraned juice vesicles)

Pith

To early explorers, it was a surprise to find that fresh citrus cured the scurvy that plagued them on long ocean voyages. Today it is no surprise to anyone that citrus is one of the most nutritious fresh fruits. Citrus is best known for its vitamin C or ascorbic acid content. While the controversy around the ability of vitamin C to cure colds is likely to continue for years, few people will deny that a healthy person is less likely to get sick.

All body tissues need vitamin C, especially connective tissues. Connective tissues make up one third of your body protein. Your skin, skeleton, muscles, blood vessels, bones and teeth are all supported, bound or held together by connective tissue. Vitamin C is essential for the formation of *colagen,* a substance that helps cement cells together. Since vitamin C aids in regeneration of connective tissue, this makes it useful in healing wounds. Vitamin C also helps in the absorption of food iron, and aids in the use of carbohydrates and amino acids. New research is linking vitamin C to many other important body functions.

Unlike most animals, man cannot manufacture his own vitamin C. It must be supplied through his diet or supplemental vitamins. The recommended daily requirement of vitamin C for an average adult male or female is 45 milligrams as determined by the National Research Council on vitamin C in 1973. The average 'Valencia' orange contains 93 mg of vitamin C. Many other valuable nutrients are found in fresh citrus. See chart on page 169.

One of the virtues of citrus is that the whole fruit is useful. To get an appreciation for the importance of this versatility, let us take a close look at a citrus fruit.

A specialist in fruit science will tell you that technically citrus fruit is called a *hesperidium.* It differs from a berry in having a leathery rind instead of solid pulp. The inner white and spongy part of the rind is known as the *albedo* or *mesocarp.* The outer colored portion is called the *flavedo* or *epicarp.* The inner segmented section, composed of membraned juice vesicles, is known as the *endocarp,* or less technically, the *pulp.* A small amount of *pith* is found in the middle.

The colorful flavedo is responsible for the striking beauty of the unblemished citrus fruit. It also has important uses. The flavedo contains many tiny oil sacs filled with distinctive and subtle flavors. The use of grated peel is one of the secrets of a gourmet cook.

These oils are also important commercially. They form the most natural citrus flavorings used in a multitude of foods. They are also ingredients in a wide array of other everyday products such as pharmaceuticals. Experiments have shown they might even be useful as plant growth regulators. It is also interesting to note that the flavedo contains more vitamin C than any other portion of the fruit.

The albedo contains a good deal of ascorbic acid or vitamin C—less than the flavedo but actually more than the pulp. It also contains *pectins*, the substance that causes jams and jellies to jell. That is why citrus is called for in many jam and jelly recipes and why the rind is used to make marmalades. Commercially, these pectins are a very important by-product.

The uses of citrus pulp are somewhat more familiar to most. Its flavor can be acid as with lemons or limes, or it can be sweet like sweet oranges or mandarins. Grapefruit is somewhat in-between. Besides the pulp itself being important for eating, the juice from citrus pulp makes one of the most popular drinks. It comes concentrated, processed, powdered, blended, reconstituted, frozen and fresh. And it's not just from oranges and lemons. Natural juices from mandarins, grapefruit and limes are becoming increasingly important. If you squeeze your own fruit, the flavor possibilities of juice blends are many.

Citrus fruit is not the only part of the tree which is commercially important. The leaves of some species contain many essential oils which are the basic ingredients of some perfumes and pharmaceuticals. Oils taken from certain citrus flowers also make some of the world's finest perfumes. Nectar from citrus flowers is loved by bees and makes excellent honey.

Purified forms of the various extracts and by-products of familiar citrus fruits on the chemist's shelf. Uses range from flavoring to coloring; perfumes to animal feeds. All of the above are processed from the peel of the fruit.

WHEN YOU'RE ADVISED TO "LIMIT SODIUM"

Think what a squeeze of fresh lemon will do. Combine it with fragrant herbs in some dishes. Use it plain for tangy tasting in others. This sunshine fruit juice is a wonderful replacement when salt is on the "no-no" list.

Think big. Don't restrict lemon juice to fruit, seafood, tomato juice and tea. Use it generously for it is very low in sodium. Use it to excite the natural flavor of almost all foods and use it as a garnish to entice the eyes and enhance the pleasure of eating. Let lemon juice and lemon wedges, strips and peels work their magic. Try fresh lemon juice in:

- Soups, such as beef or tomato bouillon, vegetable-base soups, and chowders.
- Meat, such as sirloin of beef, spread with a dry mustard, black pepper and lemon paste; steaks, chops and meatballs; meat loaf made with tomato juice and herbs, such as sweet basil or oregano; specialties, such as veal scaloppine sparked with tomato and lemon juice, rosemary and parsley.
- Salads, in marinades for vegetable and fruit combinations, such as cucumber slices and rings of onions sprinkled lightly with sugar; mixed with honey for fruit-cabbage combinations.
- Fish, rubbed inside and out with fresh lemon and dusted with dill weed before cooking, served with lemon wedges.

FRESH CITRUS NUTRITION INFORMATION

	NAVEL ORANGES	VALENCIA ORANGES	TANGERINE (DANCY)	LEMONS	1 TABLESPOON FRESH LEMON JUICE	½ GRAPEFRUIT
Food Energy	87 calories	96 calories	46 calories	24 calories	4 calories	38 calories
Protein	2.2 grams	2.3 grams	.8 grams	1.0 grams	.1 grams	.4 grams
Fat	.2 grams	.6 grams	.2 grams	.3 grams	trace	.1 grams
Carbohydrate	21.8 grams	23.4 grams	11.7 grams	7.1 grams	1.2 grams	9.9 grams
Calcium	69 mg	76 mg	40 mg	23 mg	1 mg	28 mg
Phosphorus	38 mg	42 mg	18 mg	14 mg	2 mg	17 mg
Iron	.7 mg	1.5 mg	4 mg	.5 mg	trace	.3 mg
Sodium	2 mg	2 mg	2 mg	2 mg	trace	1 mg
Potassium	333 mg	359 mg	127 mg	120 mg	21 mg	116 mg
Vitamin A	340 int'l units	380 int'l units	420 int'l units	20 int'l units	trace	10 int'l units
Thiamin	17 mg	19 mg	.06 mg	.03 mg	trace	.03 mg
Riboflavin	.07 mg	.08 mg	.02 mg	.02 mg	trace	.02 mg
Niacin	.7 mg	.8 mg	.1 mg	.1 mg	trace	.2 mg
Ascorbic Acid	105 mg	93 mg	31 mg	46 mg	7 mg	34 mg

Agricultural Handbook No. 456, Nutritive Value of American Foods. Agricultural Research Service, United States Department of Agriculture, 1975.

Ways to Peel and Slice

There probably are as many ways to peel citrus as there are citrus-lovers. On the following five pages you will find some suggestions from the "experts" for peeling, plus ideas for slicing, grating, garnishing, decorating and serving citrus. The possibilities are unlimited. It is fun to experiment so let your imagination be your guide.

Slivered orange peels are candied with sugar and honey.

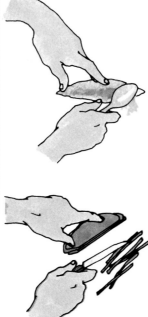

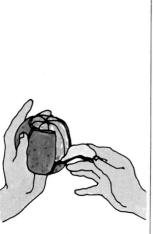

TWO EASY WAYS TO PEEL

Basketball Method:
Without cutting into the "meat," score the peel like a basketball. Pull peel away with fingers or special peeling tool.

Round and Round Method:
With a slight sawing motion, cut the peel away in a continuous spiral. For picture perfect peeled orange (ideal for bite-size pieces, cartwheels), remove all the white membrane.

Slivered Peel
Score peel of fruit into quarters; remove with fingers. With tip of spoon scrape most of white membrane from peel. Stack 2 or 3 pieces at a time on cutting board. Cut into thinnest possible strips.

Minced Peel
Prepare slivered peel as directed. Then finely chop with a knife to mince.

Orange Smiles:
"Smiles" are an appropriate name for this easy-to-eat method of preparing oranges. Slice the orange in half crosswise, then slice down across the cut into wedges. They're much easier to eat than conventional wedges because the segments pull apart.

Orange Cartwheels:
(Peeled or unpeeled). Trim a thin slice from both ends, then slice fruit crosswise. Cut slices in half for half cartwheels.

Orange Segments:
Peel fruit, using one of the methods illustrated at left. Gently separate into natural divisions.

Orange Bite-Size Pieces:
Cut peeled fruit in half lengthwise and with a shallow V-shaped cut, remove white center core. Place halves cut side down; cut lengthwise and crosswise.

Using Citrus Rinds

Citrus fruit rinds are just as useful as the juicy pulp. We already know the flavedo, or colored portion of the peel, contains tiny oil sacs with many distinctive flavors. The results of passing an orange or lemon rind over a grater just a few times can add more flavor to a recipe than the juice from one fruit.

Citrus rinds are used in many ways. Besides grating for use in recipes, thin slices can be added to drinks or frozen in ice cubes to add to drinks. Rinds are a key ingredient in marmalades; and the distinctive flavor of citron rind is a novel addition to a salad.

Citrus shells are prepared by slicing the fruit in half, gently reaming out the pulp, and then scraping the inside out with a spoon. They can be filled with fresh fruit, puddings, gelatins or whatever your imagination calls for. To prevent tipping, cut a small slice from the bottom of the fruit. See "Star Cups" page 174.

For grated peel, it is best to start with a whole fruit rather than the rind of one that has been juiced, squeezed or peeled. First wash and dry the fruit. Then, with quick, downward strokes on a grater, remove only the colored layer. For easiest handling, grate over wax paper.

Grated, sliced or shelled rinds can be placed in plastic bags and frozen. To dry grated peel, spread it thinly on a cookie sheet and bake at 200° F (93° C) for one hour. To dry sliced peel, bake at 200° F (93° C) for two hours. Store in air-tight containers.

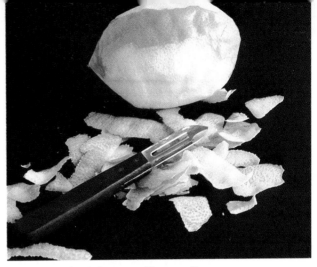
Coarse peels of citron used for appetizers.

Hand grating tool for thin slices and mincing.

Marmalade on muffins.

Marmalade

Today's marmalades are traditionally made with citrus fruits. If you are a fan of marmalades, here's one that you're sure to love—Tangerine Marmalade. It has the incomparable flavor of tangerines with just a hint of the bittersweet that you expect from marmalade. It's an ideal treat for gift giving, or to serve daily with toast, biscuits or an English muffin.

TANGERINE (MANDARIN) MARMALADE
(About seven 4-ounce jars)

2 pounds California-Arizona tangerines (approximately 6 medium tangerines)	Water to cover Sugar

Wash tangerines; cut unpeeled fruit into quarters. Separate fruit from peel and set fruit aside; cut peel into thin slices. Cover peel with water and bring to boil. Cook uncovered at medium boil until tender, about 20 to 25 minutes. Drain; measure peel to yield 3½ to 4 cups. Divide into two large saucepans. Remove seeds from tangerine segments; cut each segment into small pieces. Measure fruit to yield 3½ to 4 cups. Divide fruit into the two saucepans containing the peel. To each saucepan add as much sugar as there is peel and fruit, cup for cup. Stir well to dissolve sugar. Bring to boil and cook rapidly until jelly stage is reached. Pour into hot sterilized jars and seal at once with hot paraffin or canning jar lid.

Decorative Uses of Citrus

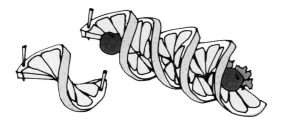

CARTWHEEL TWIST BORDER:

1. Make one cut into cartwheel from edge of peel to center.
2. Twist ends in opposite direction, standing cartwheel upright gently.
3. Use cartwheel twists to make continuous border around platter or tray.
4. Place first cartwheel twist and secure with wooden pick.
5. Continue counterclockwise to build border.
6. Slip right end of cartwheel under first cartwheel, leaving left end exposed.
7. Border can be accented with greens and reds or left plain.

BEVERAGE GARNISHES:

TIKI SAILS: Use an orange, lemon and lime cartwheel and skewer. Thread the orange first onto the skewer through the peel, the lemon next and lime last.

KABOBS:

1. Cut cartwheel into quarters. Thread two quarters onto wooden pick, placing a cherry between each.
2. Use either lemon or orange twists to thread onto wooden pick, along with olive, cherry or onion.
3. Cut off the end of an orange or lemon, that has been fluted diagonally. Place on end of skewer with cherry.

SCALLOPED BORDER FOR TRAYS, PLATTERS OR CARVING BOARDS:

1. Place half-cartwheels of orange or grapefruit on outer edge of tray to be garnished end to end.
2. Place second layer of smaller half-cartwheels (lemon) on top of first layer, aligning cut edge.
3. Place third layer of still smaller half-cartwheels (lime) last. The result is a scalloped, layered effect.

CARTWHEEL COMBINATION BORDER:

Another option is a simple combination of plain and fluted cartwheels overlapping to make a border.

DECORATED CARTWHEELS

Use any one of the following to decorate plain or fluted cartwheels:

Pimento cut-outs	Strawberry twist	Lemon and lime twist	Lemon and pimento

Lime twist	Olive and parsley	Paprika or parsley	Cookie	Cloves

Pickle fan

FLUTED CARTWHEELS:

Best tool for peeling is called a zester (or cesteur) which can be found at gourmet shops.
1. Hold stem end of fruit with thumb and middle finger.
2. Use twist maker and peel from end to end leaving about ¼ to ½ inch between each cut.
3. Use cartwheels to desired thickness.

CHRYSANTHEMUM FLOWER:

1. Use grapefruit, orange, lemon and lime. Be sure peel is clean.
2. Use star-cut method shown in Star Garnish on the next page. Make "jaws" with long sawtooth cuts, cutting teeth to within ¾ inch of ends.
3. Separate.
4. Remove pulp from petals with fingers, saving pulp for use in fruit cup.
5. Take pointed wooden skewer and place two grapefruit flowers on it. Add two orange flowers, then a lemon and a lime.
6. Cut wooden skewer "stem" of flower to length needed. If flower is not to be used right away, place in plastic to keep fresh.
7. Surround flower with watercress or other greens.

STAR GARNISH:

1. Hold fruit with thumb and middle finger at stem and blossom ends.
2. Make "jaws" with a sawtooth-cut around middle, cutting to center of fruit. Cut through each time to allow clean separation.
3. Using both hands, gently pull fruit apart. If it does not pull apart cut through still uncut skin, then separate.

STAR CUPS:

1. Use an orange or grapefruit.
2. Make sawtooth-cut finer and not as deep as for the chrysanthemum.
3. Using a grapefruit knife, clean out pulp, leaving enough next to peel for color.
4. Fill cups with berries, sherbet, puddings, relishes or whatever you wish.

CITRUS ROSES:

Roses can be made from grapefruit, oranges, lemons and mandarins. Western-grown grapefruit have slightly thicker peel, letting you make two roses from one fruit. After cutting off thin outer peel, cut inner white membrane.

1. Peel the outer skin in a wide, thin, continuous spiral.
2. Place the peel in near-boiling water for one or two minutes to make it more flexible. Cool with cool water to handle easily.
3. To make rose, wind peel in reverse with colored side in and the white side out.
4. Starting with the center of the rose, form the peel tightly into a bud. Secure with wooden pick at base.
5. Continue forming flowers by keeping petals next to stem tight, letting outer layers be looser. Secure flower at base with wooden pick as rose is formed.
6. Place in ice water to set flower.
7. Cut off visible part of wooden picks before placing in arrangement.

INDEX

A
Acclimating citrus, 144
Albedo, 168
'American Wonder' lemon, see 'Ponderosa Lemon'
Aphids, 134
Arizona sweets, 47
Arizona, 14, 17, 59, 66, 67, 87, 91
Atlanta, 6
Atlas, 6

B
Balled and burlapped (B&B),121
Bark, 117
Bartender's lime, see 'Mexican' lime
Basin, 123
Basketball peeling method, 170
'Bearss' lime, 94, 96, 141, 146, 162
Beverage garnishes, 172
Bitters, William, 17
Black scale, 134
Blood meal, 142
Blood orange, 9, 62-65, 130
  'Moro', 65
  'Sanguinelli', 64
  'Tarocco', 63
Boron, 126
'Bonpeiyn' pummelo, 97
'Bouquet de Fleurs' sour orange, 9, 59
'Bouquet' sour orange, 61, 118, 151, 152
Brown scale, 134
Bud union, 117, 118
'Buddha's Hand' citron, 7, 10, 114, 155

C
Calamondin, 111, 113, 130, 139, 156
Calcium carbonate lime, 142
Calcium, 125, 142
California, 12, 13, 14, 17, 27, 59, 66, 67, 82, 87, 91, 96, 147
California red scale, 134
Cartwheel combination border, 172
Cartwheel twist border, 172
Castillo, Bernard Diaz del, 8, 11
'Chandler' pummelo, 97, 99, 100, 118
'Changsha' tangerine, 139
Chinan, 108
Chinese box orange, 156
'Chinese New Year' orange, 141, see 'Otaheite' orange
'Chinotto' sour orange, 59, 60, 118, 138, 153
Chrysanthemum flower garnish, 173
Citrangequat, 111, 113
Citron, 7, 11, 114-115, 154
  'Buddha's Hand', 7, 10, 114, 115, 155
  'Etrog', 115
  'Fingered', 115
Citrus and religion, 7
Citrus and the conquerors, 8
Citrus blast, 133
Citrus blossoms, 117
Citrus climates, 25
Citrus cooperatives, 15
Citrus diseases, 132
Citrus hedge, 159
Citrus history, 10
Citrus indoors, 141
Citrus industry today, 17
Citrus in containers, 160
Citrus in northern California, 15
Citrus in Texas, 136
Citrus in the lawn, 124
Citrus in the west, 14
Citrus in tropical climates, 25
Citrus legend continues, 17
Citrus red mites, 135
Citrus rinds, 171
Citrus roses, (garnishes) 174
Citrus shrubs, 160
Citrus spoilers, 132
Citrus thrips, 135
Citrus travels, 9
Citrus: yesterday and today, 5
Clausena lansium, 158
Clay soil, 120

'Clementine' mandarin, 69, 70, 106, 138
Climates, 27
Cold tolerance, 119
Common oranges, 47
Containers, 120, 127, 160
Container soils, 120
Copper, 125
Cornell peat-lite mix, 142
Cortez, 8
Cottony cushion scale, 134
Cuban shaddock, 131
Culls, 20
'Cunningham' citrange, 131

D
'Dancy' mandarin, 69, 71, 102, 117, 118, 146
Decorated cartwheels, 173
Decorative uses of citrus, 172-174
Dictyospermum scale, 134
'Diller' orange, 47, 58
Dillon, Floyd, 16, 17
Dolomite lime, 142
Drip hose, 123
Drip line, 123
Dry root rot, 133
du Pont, 144
'Duncan' grapefruit, 88, 91, 102
Dwarf citrus, 16, 131
Dwarf tree, 121
'Dweet' tangor, 106, 107

E
'Encore' mandarin, 69, 72
Endocarp, 168
Enjoying citrus, 163
Espaliers, 161
'Etrog' citron, 7, 114, 115
'Eureka' lemon, 83, 84, 87, 118, 130, 146
European brown snails, 135
'Eustis' limequat, 118, 139

F
'Fairchild' mandarin, 69, 73, 138, 163
Fertilizer, 116, 124, 126, 127, 143
Fertilizing container citrus, 127
'Fingered' citron, 7, 115, 155
Fingerlime, 156
First California orange grove, 14
Flavedo, 168
Florida, 12, 13, 27, 66, 67, 91, 96, 147
Florida red scale, 134
Flowers and fruit, 119, 143
Fluted cartwheels, 173
'Fortune', 69
Fortunella, 157
From tree to market, 18
Frost damage, 129
Frozen leaves, 129
Fruit drop, 133
Fruit split, 133
Fruit, 119

G
Golden apples, 6
Good oranges, 67
Gophers, 133
Gorgon, 6
Grading, 21
Graft union, 117, 118
Grapefruit, 88-91, 130, 139, 147, 166
  'Duncan', 88, 91, 102
  'Marsh Seedless', 89
  'Redblush', 90
  'Ruby', see 'Redblush'
  'Star Ruby', see 'Redblush'
  'Thompson', see 'Redblush'
Grapefruit at the supermarket, 91
Ground bark, 142
Grouping plants, 145
Growing citrus, 117
Growing citrus in Texas, 136-139
Growing citrus outside the citrus belt, 141
Growing dwarf citrus, 130
Growing pains, 15,
Gumming, 133
Gummosis or foot rot, 132

H
Hadar, 7, 114
'Hamlin' orange, 47, 55, 67, 138
Hand picked, 18
Hippomenes, 7
'Honey' mandarin, 69, 72, 80
Hoof and horn meal, 142
How to peel a pummelo, 98-99
How to fertilize, 127
Humidity, 144, 145

I
'Improved Meyer' lemon, 83, 86
Indoor citrus, 141
Indoor citrus society, 147
Indoor citrus questionnaire, 146
Iron, 125
Iron deficiency, 125
Iron sulfate, 142
Isles of Hesperides, 6

J
'Jaffa' orange, 9, 47, 67
Juices, 23
Juicing fresh citrus, 164
June drop, 119
Juno, 6
Jupiter, 6

K
Kabobs, 172
'Kao Pan' pummelo, 97
'Kao Phaung' pummelo, 97
'Kara' mandarin, 69, 74, 80, 118, 146
'Key' lime, 92
'King' mandarin, 69, 106
'Kinnow' mandarin, 69, 75, 80
Kino, Father Eusebio, 14
Kumquat, 108-110, 130, 139, 147, 157
  'Meiwa', 109
  'Nagami', 110
Kumquat Hybrid, 111-113
  'Calamondin', 113
  'Citrangequat', 113
  'Limequat', 112
  'Orangequat', 112

L
'Lake', 80
Landscaping with citrus, 149
Leaf burn, 145
Leaf drop, 133
Leaves, 118
Lemon, 10, 11, 82-87, 119, 128, 147, 166
  'Eureka', 84
  'Improved Meyer', 86
  'Lisbon', 85
  'Meyer', 86
  'Ponderosa', 85
Lemonade, 11
Lemons at the supermarket, 87
Lewis, Judge Joseph, 15
Lime, 11, 92-96, 147
  'Bearss', 94
  'Key', 92
  'Mexican', 93
  'Persian', 9, 92, 96, 141, 146, 147
  'Rangpur', 95
  'Tahitian', 9, 92
Limeberry, 157
Limequat, 111, 112, 130
Limes at the supermarket, 96
Limestone, 142
'Lisbon' lemon, 82, 85, 87, 130
Loam soil, 120
Louis XIV, 6, 7, 150
Louis XVI, 6

M
Magnesium, 124
Mail order citrus, 146-147
Main scaffolds, 129
Maintenance pruning, 129
Mandarin, 69-81, 106, 130, 141, 165
  'Clementine', 70
  'Dancy', 71
  'Encore', 72
  'Fairchild', 73
  'Honey', 72

INDEX

'Kara', 74
'King', 69, 106
'Kinnow', 75
'Mediterranean', 77
'Murcott', 80
'Page', 76
'Pixie', 77
'Satsuma', 78
'Wilking', 79
Mandarins at the supermarket, 80
Manganese, 125
'Marrs' orange, 47, 56, 137, 138
'Marsh Seedless' grapefruit, 88, 89, 91, 118, 130, 137, 144, 146, 162
'Marsh White' grapefruit, 139
'Mato Butan' pummelo, 97
Mealybugs, 134
Median apple, 10
'Mediterranean' mandarin, 77, 118
'Meiwa' kumquat, 108, 109
Mesocarp, 168
'Mexican' lime, 92, 93, 96, 139, 141
'Meyer' lemon, 82, 83, 118, 131, 139, 141, 162, see also 'Improved Meyer' lemon
Microclimates, 32
Micronutrients, 128
Minced peel, 170
Mini-greenhouse, 145
'Minneola' tangelo, 80, 102, 103, 130, 131
Misting, 145
'Mitis' dwarf orange, 147
'Moanlua' pummelo, 97
Molybdenum, 126
'Moro' blood orange, 62, 65, 162
Moses, 7
Mother orange, 15
Mound planting, 120
Mulch, 124
Murcott orange, 80
'Murcott' mandarin, 80
'Myrtifolia' oranges, 153

N 'Nagami' kumquat, 108, 110, 118, 146
Navel oranges, 20, 47, 130, 138
'Nippon' orangequat, 118
Nitrogen deficiency, 125
Nitrogen, 124
Nucellar embryony, 117
Nutrients, 124
Nutrition information, 169

O Oak root fungus, 132
Oil glands, 168
Orange, 45-65, 165
    Blood, 62-65
    Sour, 59-61
    Sweet, 45-58
Orange bite-size pieces, 170
Orange cartwheels, 170
Orange jessamine, 158
Orange juice, 22, 164-165
Orange segments, 170
Orange smiles, 170
Orangequat, 111, 112, 130
Orangerie, 5, 6
Oranges at the supermarket, 66-67
'Orlando' tangelo, 80, 102, 104, 138
'Otaheite' orange, 141, 146, 147, 155
'Owari' satsuma mandarins, 118, 146

P Packing house, 20
'Page' mandarin, 76, 162
'Paradise' grapefruit, 146
'Parson Brown' orange, 47, 67
Parthenocarpy, 119
Peat moss, 142
Pebble trays, 145
Peel and slice, 170
Perlite, 142
Perseus, 6
Persian apple, 10
'Persian' lime, 9, 92, 96, 141, 146, 147

Pests, 134, 145
Phosphorous, 124
'Pineapple' orange, 47, 57, 67
Pith, 168
'Pixie' mandarin, 77
Planting step by step, 122
Planting, 122
Pollination, 119, 143
Poncirus trifoliata, 157
'Ponderosa' lemon, 82, 83, 85, 130, 131, 141, 146, 147
Potasssium, 124
Potassium nitrate, 142
Potassium sulfate, 142
Potting soil, 121
Preparation of planting hole, 122
Principal parts of orange, 168
Protect yourself, 129·
Protection of severe cut, 129
Pruning, 128
Pruning frost damaged trees, 129
Pruning young trees, 129
Pulp, 168
Pummelo, 7, 10, 97-101
    'Bonpeiyn', 97
    'Chandler', 100
    How to peel, 98-99
    'Kao Pan', 97
    'Kao Phang', 97
    'Mato Bhutan', 97
    'Moanlua', 97
    'Reinking', 101
    'Siamese Pink', 97
Purple scale, 134

Q Quick decline, 133

R Raised bed, 120
'Rangpur' lime, 95, 141, 155
Rare Fruit Council, 139
'Redblush' grapefruit, 88, 90, 91, 137, 139, 146
'Reinking' pummelo, 97, 98, 101
Rejuvenation pruning, 128
Rind, 168
Rio Grande Valley, 137
'Robertson' navel orange, 46, 47, 51, 146
Roots, 116, 118
Rots, 22
Round and Round peeling method, 170
'Royal' mandarin, see Tangor
'Ruby Red' grapefruit, 139, see 'Redblush'
'Ruby' grapefruit, 88, 91

S 'Sampson' tangelo, 102, 105
Sand, 142
Sandy soil, 120
'Sanguinelli' blood orange, 9, 64, 118
Satsuma mandarin, 69, 78, 138
Scales, 134
Scalloped border, 172
Scion, 118
Seediness, 119
Seedlings, 118
Seeds, 117
Selecting good oranges, 67
Selecting trees, 121
'Seminole' tangelo, 80, 81, 105
Serra, Father Junipero, 14
'Seville' sour orange, 9, 59, 61
'Shamouti' orange, 9, 47, 54, 67
'Siamese Pink' pummelo, 97
'Sinton' citrangequat, 138, 158
'Skaggs Bonanza' orange, 47, 50
Skirt picking, 18
Sleeves, 122
Slivered peel, 170
Sodium, 169
Soil, 116, 142
Soil type, 120
Sour orange, 59-61, 138, 153
    'Bouquet', 61
    'Chinotto', 60
    'Seville', 61

Specialty trees, 138
Sphagnum peat moss, 142, 145
Splits, 22
Standard tree, 121
Star roses, 174
'Star Ruby' grapefruit, 91, 137, 139
Star cups, 174
Star garnish, 174
Storing fresh juice, 167
Sulfur, 125
'Summernavel' orange, 52
Sunburn prevention, 129
Sunburn, 133
Sunkist growers, 13, 16, 17, 20
Superphosphate, 142
Supplemental light, 143
Surinam, 27
Sweet orange, 45-58
    'Diller', 58
    'Hamlin', 55
    'Marrs', 56
    'Pineapple', 57
    'Robertson' Navel, 51
    'Shamouti', 54
    'Skaggs Bonanza', 50
    'Summernavel', 52
    'Trovita', 49
    'Valencia', 53
    'Washington', 48

T 'Tahiti' orange, 141, 147, see 'Otaheite' orange
'Tahitian' lime, 9, 92
Tangerine, 69, see Mandarin
Tangelo, 80, 102-105, 130
    'Minneola', 103
    'Orlando', 104
    'Sampson', 105
    'Seminole', 106
Tangerines and tangelos, 138
Tangor, 69, 81, 106-107
    'Dweet', 106, 107
    'Temple', 106, 107
'Tarocco' blood orange, 63
'Temple' tangor, 81, 106, 107
Texas, 13, 17, 91
Texas climate, 137
'Thomasville' citrangequat, 138
'Thompson' grapefruit, 91, see 'Redblush'
Tibbets, Eliza, 14
Tiki sails, 172
Tree selection, 121
Trifoliate orange, 139, 157
Triphasia trifolia, 157
Tristeza, 133
'Trovita' orange, 46,47, 49
'Tule Gold' navel orange, 47

U University of California soilless mixes, 142
Urea formaldehyde, 142

V 'Valencia' orange, 9, 21, 46, 47, 53, 66, 67, 130, 137, 138, 162
Variegated calamondin, 118
Vermiculite, 142
Virus disease, 133
Vitamin C, 168

W Wampee, 158
'Washington' navel orange, 12, 13, 14, 46, 47, 48, 67, 118, 119, 146, 163
Water, 116, 142
Watering, 123
Weather protection, 122
Where to plant, 122
Which citrus?, 159
Which variety?, 137
'Wilking' mandarin, 69, 79
'Willowleaf', 69
Wolfskill, William, 14

Y Yellow scale, 134

Z Zinc, 125
Zinc deficiency, 125